FISHING IN CONTESTED WATERS

Place and Community in Burnt Church/Esgenoôpetitj

After the Supreme Court of Canada's 1999 *Marshall* decision recognized Mi'kmaw fishers' treaty right to fish, the fishers entered the inshore lobster fishery across Atlantic Canada. At Burnt Church/Esgenoôpetitj, New Brunswick, the Mi'kmaw fishery provoked violent confrontations with neighbours and the Canadian government. Over the next two years, boats, cottages, and a sacred grove were burned, people were shot at and beaten, boats rammed and sunk, roads barricaded, and the local wharf occupied.

Based on 12 months of ethnographic field work in Burnt Church/Esgenoôpetitj, *Fishing in Contested Waters* explores the origins of this dispute and the beliefs and experiences that motivated the locals involved in it. Weaving the perspectives of native and non-native people together, Sarah King examines the community as a contested place, simultaneously Mi'kmaw and Canadian. Drawing on philosophy and indigenous, environmental, and religious studies, *Fishing in Contested Waters* demonstrates the deep roots of contemporary conflicts over rights, sovereignty, conservation, and identity.

SARAH J. KING is an assistant professor in the Liberal Studies Department at Grand Valley State University.

T0256846

Fishing in Contested Waters

Place and Community in
Burnt Church/Esgenoôpetitj

SARAH J. KING

UNIVERSITY OF TORONTO PRESS
Toronto Buffalo London

© University of Toronto Press 2014
Toronto Buffalo London
www.utppublishing.com
Printed in Canada

ISBN 978-1-4426-4176-1 (cloth)
ISBN 978-1-4426-1096-5 (paper)

Library and Archives Canada Cataloguing in Publication

King, Sarah J. (Sarah Jean), 1973–, author
Fishing in contested waters : place and community in Burnt Church / Esgenoôpetitj / Sarah J. King.

Based on thesis (doctoral) – University of Toronto, 2008, under title: Contested place: religion and values in the dispute, Burnt Church / Esgenoopetitj, New Brunswick.

Includes bibliographical references and index.
ISBN 978-1-4426-4176-1 (bound). ISBN 978-1-4426-1096-5 (pbk.)

1. Micmac Indians – Fishing – New Brunswick – Burnt Church. 2. Micmac Indians – New Brunswick – Claims. 3. Micmac Indians – New Brunswick – Government relations. 4. Lobster industry – Social aspects – New Brunswick – Burnt Church. 5. Culture conflict – New Brunswick – Burnt Church. 6. Burnt Church (N.B.) – Ethnic relations. I. Title.

E99.M6K55 2013   304.208997'343071521   C2013-904614-3

This book has been published with the help of a grant from the Canadian Federation for the Humanities and Social Sciences, through the Awards to Scholarly Publications Program, using funds provided by the Social Sciences and Humanities Research Council of Canada.

This book has been published with the support of a Book Award from Wilfrid Laurier University.

University of Toronto Press acknowledges the financial assistance to its publishing program of the Canada Council for the Arts and the Ontario Arts Council.

University of Toronto Press acknowledges the financial support of the Government of Canada through the Canada Book Fund for its publishing activities.

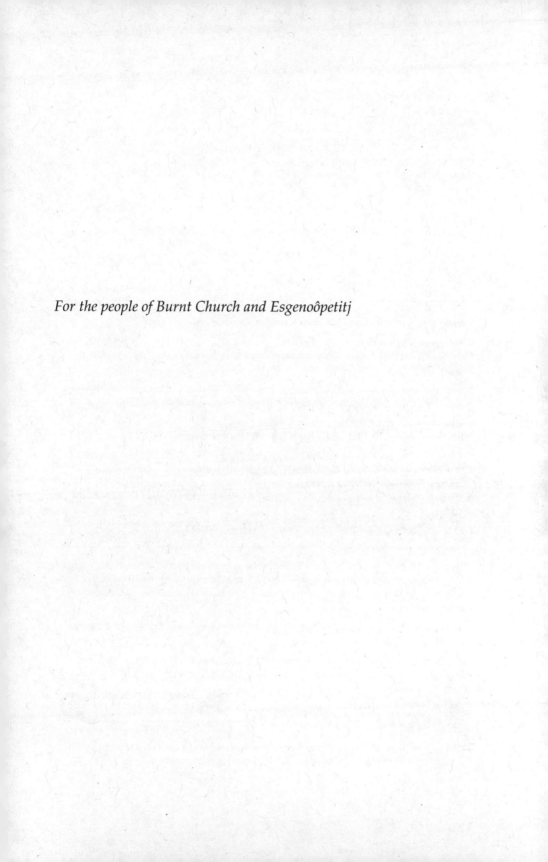

For the people of Burnt Church and Esgenoôpetitj

Contents

Figures

Acknowledgments

This book only exists because of the people in Burnt Church and Esgenoôpetitj who welcomed me when I arrived in their community asking about the dispute. I want to honour each of these people for the guidance, insight, challenge, and support they gave to me, in spite of the possible risk to themselves. Thank you.

I'd also like to recognize the religious communities and activist groups who welcomed this research, and who shared their experiences and insights with me. Their stories of the dispute in Burnt Church and Esgenoôpetitj have also informed this book, and I thank them.

Some of the people and groups who participated in this research are mentioned by name in this book, but many are not named in order to protect their identities. I've not named anyone here in these words of thanks, because to name some would risk valuing their insights more than those of others who are unnamed. Many people made tremendously generous contributions by telling their stories, inviting me into their homes, and including me in community events, by sharing their archives, by introducing me to others within their community, and by questioning and challenging me. Thank you to all of you.

If you'd like to be in touch with me about any of this (and as long as there's still email in the world), you can contact me at dr.sarah.king@gmail.com.

Most of the stories in this book aren't mine – they are stories shared with me by the people of Burnt Church and the Esgenoôpetitj First Nation. This book builds upon and engages these stories in ways that are entirely my own. My hope is that I have accurately reflected local people's experiences and clearly distinguished them from my own analysis and opinions. All mistakes are my responsibility.

I want to thank my colleagues at the University of Toronto Centre for Environment and Centre for the Study of Religion for their advice and support. Ingrid Stefanovic, my friend and mentor, has been unfailingly generous in her support of this project. Stephen Bede Scharper and Hilary Cunningham provided key guidance in the early stages; thanks to them, to Deborah McGregor, Michael Ostling, David Perley, Laurel Zwissler, and Kenn Maly, and to Robert Mugerauer of the University of Washington Seattle.

My colleagues at Wilfrid Laurier University and Queen's University also have my deep thanks, including (but not limited to) Carol Duncan, Mick Smith, Michel Desjardins, Janet McLellan, Kay Koppedrayer, Ron Grimes, Chris Klassen, Mona Lafosse, Brian Cumming, Melissa Ireland, and Lauren Price. My colleagues at Grand Valley State University have been welcoming and kind as I finished this book while settling in. Thanks also to those who engaged this project as it developed, including Janet McLellan and her graduate methods class at WLU (particularly Yasaman S. Munro); the Philosophy and Religion department at Central Michigan University; Magda Kazubowski-Houston, Whitney Lackenbauer, Yale Belanger, Marc Fonda, Nick Shrubsole, and my colleagues at numerous academic conferences; and students in the three iterations of my course "Religion and Colonialism in Canada" at WLU.

At the University of Toronto Press my editors Virgil Duff and Douglas Hildebrand have been patient and persistent on my behalf. Thanks to them, and to the anonymous reviewers who offered tremendous careful insight as the book developed.

The research for this book was undertaken with the support of a Social Sciences and Humanities Research Council of Canada Doctoral Fellowship, as well as the Arthur and Sonia Labatt Fellowship in Environmental Studies, Edward W. Nuffield Travel Grant, and Thomas and Beverley Simpson Ontario Graduate Scholarship at the University of Toronto.

Finally, writing this book has taken me on an important personal journey, as well as a professional one. Thanks to my friends and family who have walked with me and stuck with me.

Preface

This book presents the story of a conflict amongst Canadian fisheries communities that captured global attention from 1999 to 2002. In Burnt Church, New Brunswick, members of the Mi'kmaw community entered the lobster fishery in the face of violent opposition from commercial fishers and from the Canadian government, which regulates the fishery. The Mi'kmaw fishery began after the Supreme Court of Canada handed down the *Marshall* decision, recognizing that the Mi'kmaq had a right to fish according to the Treaties of Peace and Friendship (1760/61). The court found that current fisheries legislation was unconstitutional, as it restricted the Mi'kmaq ability to fish and sell fish. In the Mi'kmaw community, after the court's decision, people celebrated: "It was like time slowed down. That's how intense and wonderful it was." But when the Mi'kmaq at Burnt Church began to fish – without first obtaining licences from the federal government – happiness turned to violent conflict.

Non-native neighbours were convinced that the Mi'kmaw fishery would decimate commercial lobster stocks; the Canadian government was concerned about the integrity of the stocks and also of their system of fisheries management. Violence broke out between the Mi'kmaq and non-native commercial fishers. Trucks and buildings were burnt, people were beaten with baseball bats, and some began to arm themselves. As the RCMP and the Department of Fisheries and Oceans stepped in, their dealings with the Mi'kmaq also became violent. Video images of government boats ramming and capsizing Mi'kmaw fishing dories drew attention from around the world, as did images of the Mi'kmaw occupation of the Burnt Church wharf. People in the Burnt Church First Nation were risking their lives in order to continue their fishery. Why?

One reason is the defining historical importance of the fishery to two groups of people with an intricate and complex relationship. Two communities share the name Burnt Church. The Burnt Church First Nation, known in Mi'kmaq as Esgenoôpetitj, is situated on the shores of Miramichi Bay in northeastern New Brunswick. Immediately beside the First Nation is a small village of English-speaking settlers, also known as Burnt Church. The two communities of Burnt Church, Mi'kmaw and English, exist side by side but are very separate places. They are neighbours of long standing and know one another very well, yet not at all. This peculiar juxtaposition of closeness and estrangement raises questions about relationships between indigenous people and settlers both in Burnt Church and across the broader Canadian context.

Though the dispute centred around the lobster fishery, fisheries management in Burnt Church was not the only issue. The dispute was one moment in the ongoing contestation of place that has been happening in Burnt Church/Esgenoôpetitj – and in Canada – since the arrival of the first settlers. The activism of the dispute represented, for the Mi'kmaq, an exercise of their rights and sovereignty in their own lands and waters as affirmed in the treaties and by Canadian courts. For the residents of the English village of Burnt Church, the dispute threatened not only their livelihoods but their identities as members of the Canadian nation. At its heart, this is a book about experiences and interpretations of the dispute in the two communities that call themselves Burnt Church. It is based upon fieldwork completed in these communities shortly after the dispute subsided. The stories of people who participated in the dispute, and those who lived through it, provide the central lens through which the dispute is understood. This is also a work about the construction and contestation of place in Burnt Church and Esgenoôpetitj, and an attempt to grapple with the values that shape colonial Canada.

This book has its beginnings in the events of the dispute in Burnt Church, and in the theoretical and methodological approaches of its author. Chapters 1 and 2 engage these fundamental frameworks, and are the foundation of the thematic explorations that come in later chapters. Chapter 1 discusses the importance of prioritizing lived experience in understanding colonial contexts, and what phenomenological and ethnographic approaches can offer to an understanding of the dispute. It explores the nature and significance of place, and the ethical challenges of ethnographic fieldwork in this particular place. Chapter 2 is a detailed account of the events of the dispute, told from the perspec-

tives of people in the Mi'kmaw and English communities. It is an attempt to share the story of the dispute as it was experienced by local
people who were directly engaged, willingly or reluctantly, in a very
public conflict.

Chapters 3 through 5 explore the values and ideas that the English
and Mi'kmaw residents of Burnt Church use to frame their motivations
in and experiences of the dispute. Chapter 3 outlines the cultural, religious, and historic construction of Burnt Church/Esgenoôpetitj as a
contested place. The two communities of Burnt Church have very separate notions and experiences of place, though they inhabit the same
landscape. Each community is very specifically and richly tied to its
place, and yet, for both, the threat of displacement has become an integral part of place itself. At some level, for each community, the very
presence of the other Burnt Church is a reminder of the history or possibility of displacement. Each community has developed a sense of its
place as contested, where the right of continued residence, authority,
and belonging is constantly challenged and defended.

Chapter 4 explores the importance of rights, sovereignty, and nationalism for people in the dispute. It contextualizes the different understandings of these values within each community, and characterizes
their significance through stories about daily life in the dispute from
community members. In Esgenoôpetitj, people's concerns for indigenous rights and Mi'kmaw sovereignty motivated their perseverance in
the dispute, and continued to be important issues for many residents
after the dispute subsided. In the English village, the experience of the
dispute raised questions and concerns for many people about their
place as Canadians and their relationship to the Canadian government.
In both communities, issues of rights, sovereignty, and nationalism
were not simply political issues, but were complex problems of world
view, religion, and identity that engaged individuals, groups, and
sometimes the entire community. The prolonged conflict led people in
both communities to re-examine their identities, as Mi'kmaq and/or
as Canadians.

Chapter 5 explores the central role of conservation discourse in Burnt
Church and Esgenoôpetitj, both as a restriction of and an opportunity
for the presentation of local concerns. The sustainable management of
lobster stocks was the Canadian government's stated priority at Burnt
Church, and the sole focus of mediation and negotiation. All parties
positioned themselves as the most able to manage fishery resources according to broad conservationist principles. This focus on conservation

as the centre of the dispute left little room to address other concerns of the communities of Burnt Church (such as sovereignty or livelihood), unless these were presented as dimensions of conservation. During the dispute, the discourse of conservation became the framework within which the concerns, values, and aspirations of each of the Burnt Churches were presented – their concerns for livelihood, critiques of the government and negotiations for alliances. The imposition of the discourse of conservation in Burnt Church is a dimension of the larger phenomenon of the colonization and globalization of indigenous and rural communities, as Guha (1989) and Vandergeest and DuPuis (1996) have outlined. At the same time, conservation became a critical way in which both communities articulated and defended their positions, and their places.

Chapter 6 examines the importance and impact of non-local conceptions of place in the dispute, such as the views and ideas of governments and activists who were involved in the conflict. I argue that local perception of the Canadian government's approach to the dispute as calculative and managerial became a major factor in the ongoing violence. The government's unwillingness to recognize the multiple issues and understandings under dispute prolonged and intensified the conflict. The government justified its actions through appeals to order and authority, and in the language of nationalist myth. These myths are potent for other players in the dispute, including non-native activists. The chapter concludes with an exploration of the nature and challenges of the colonial multiplicity that is Canadian reality.

Resource access and management are not simply scientific, legal, or economic problems. They are complex conundrums that arise because many groups, communities, or nations can all have unique ties to a specific place simultaneously. Burnt Church is a Mi'kmaw place, Esgenoôpetitj. It is also a settled place, home to non-native Canadians. This book explores the values and experiences which have shaped, and are shaped by, this contested place. Understanding why the dispute happened in Burnt Church is important to the communities of Burnt Church and Esgenoôpetitj, and also in the larger Canadian, colonial, and globalizing contexts.

FISHING IN CONTESTED WATERS

Place and Community in Burnt Church/Esgenoôpetitj

1 Introduction: Re-membering Burnt Church

In November 1998, the Supreme Court of Canada heard an appeal in the case of Donald Marshall Jr. Marshall, a Mi'kmaw man from Nova Scotia, was charged under federal fishery regulations with fishing and selling eels illegally.[1] Marshall admitted that he had been fishing eels, and argued that the Mi'kmaq held the right to fish according to the peace and friendship treaties they signed with the British in 1760–1. Therefore, he argued, the regulations of the Canadian government did not apply to Mi'kmaw fishers.

In a decision handed down on 17 September 1999, the court recognized Marshall's treaty right to fish, sell fish, and gather wildlife, saying "nothing less would uphold the honour and integrity of the Crown in its dealings with the Mi'kmaq people to secure their peace and friendship, as best the content of those treaty promises can now be ascertained."[2] The court ruled that existing fishing regulations were unconstitutional, as they restricted these rights without justification. Within days, Mi'kmaw people began to enter the lobster fishery across Atlantic Canada.

In the community of Esgenoôpetitj (the Burnt Church First Nation) the court's decision was greeted with great celebration. Previous rulings in favour of native rights to harvest and sell natural resources, such as timber, had resulted in little new employment under the terms of agreements signed by the elected chief and the federal government. In 1999, the community had just held an election in which the long-standing chief, Wilbur Dedam, and his allies had lost their majority on council. In this shifting political landscape, the *Marshall* decision was seen as an opportunity for people to exercise their treaty rights on their own terms, in order to make a living. As one couple said, "When the

Figure 1. Atlantic Canada and Mi'kma'ki. Burnt Church/ Esgenoôpetitj is located in northeastern New Brunswick, in the northern portion of Mi'kma'ki (traditional Mi'kmaw territory). (Source: The GIS and Cartography Office, Department of Geography, University of Toronto)

Marshall decision came down, everybody just said 'All right! We have fishery rights now!' And they didn't even consult with the council or anything, they just started putting their traps in – bang bang bang bang … It was pretty exciting, really. People were happy" (Dalton and Cindy).[3] The sense of celebration in the community continued, as more and more people began to fish. People worked cooperatively in small fishing dories, purchased refurbished traps, and hauled in their catch by hand. According to locals, "You could see, down the far flung reaches of our community, our people were so excited about the possibility of permanent employment" (Miigam'agan).

The non-native communities that neighbour Esgenoôpetitj rely heavily on the lobster fishery. In the English village of Burnt Church, almost every resident either fishes lobster or has family members who do. The growing native fishery caused increasing concern in the English village. In the words of one resident,

> Almost within a week, we had Indians from Big Cove, from Millbrook and from the Burnt Church Reserve all converged on the wharf over here with their fishing boats, and they started fishing lobster in September. There has never been a fishery out here in September; it has always been a spring fishery. So all of a sudden there were a tremendous amount of people fishing here. And that raised a great deal of concern from the people living here, that fished here for years ... If they're out there destroying the stock then they're gonna ruin the fishery. (Paul)[4]

The Mi'kmaw fishery did not conform to the federal government's regulated season.[5] Native fishers across the region were fishing without licences and tags, which is also against government regulation. Concerns for the stock, and for the unfairness of a situation where some people were perceived to be acting outside the law, were escalating rapidly.

Non-native commercial fishers and their families organized a protest, concerned with the lack of government enforcement action against Mi'kmaw fishers. Violence erupted, on both sides. Native fishing equipment was hauled up and destroyed. Trucks and cottages were burnt in the non-native community, and the sacred arbour on the reserve was torched. People threatened each others' lives, beat others with baseball bats, and shot at one another. This violence did not deter the Mi'kmaq of Esgenoôpetitj from fishing, which they continued to do according to the treaties which had been recognized by the Marshall decision.

Yet, while the Mi'kmaq believed that they had the right to fish under their own system of governance, the Government of Canada did not. The RCMP, the Department of Fisheries and Oceans (DFO), and the Coast Guard came to Burnt Church to enforce Canadian fisheries regulations and became participants in the violence. Native boats were rammed and capsized by government vessels, and native activists were beaten and threatened. Surveillance helicopters and RCMP command posts became a normal presence in Burnt Church.

The Burnt Church First Nation is situated on the shores of Miramichi Bay in northeastern New Brunswick. Immediately beside the First Nation is a small village of English settlers, also known as Burnt Church. Their neighbours to the northeast are francophones who settled the "Acadian Peninsula" after their expulsion from Nova Scotia in 1755. To the southwest lies the region known as the Miramichi, historically dominated by the great river system that gives the region its name. The neighbouring settler communities there are largely, but not entirely, anglophone. The Burnt Church First Nation is one of many Mi'kmaw communities across the region, which is part of the traditional territory of the Mi'kmaq, Mi'kma'ki.[6] The two communities of Burnt Church, Mi'kmaw and English, exist side by side on the shores of Miramichi Bay, but they remain very separate places. Although the fishing dispute in Burnt Church involved many in the region and across the country, it happened on the waters and lands of these two communities – a shared experience that reflected and continued the deep divisions that exist between them.

The dispute, as the locals call it, lasted more than three years. In 2001, elected Chief Dedam was returned his majority on the band council. In August of 2002, he signed an agreement-in-principle with the Government of Canada, accepting Canadian governance of the Mi'kmaw fishery in return for money, boats, and licences. The conflict subsided, and Burnt Church slipped out of the news.

But questions remain. What motivated people to risk their safety, and even their lives, over this conflict? What were the roots of this violence between people who had been neighbours for generations? And why was it that the Canadian government participated in largely inflaming the problem rather than resolving it? As the dispute drew to a close, a senior official in the DFO commented, "Perhaps it never really was about fish" (CBC 2002). Indeed.

This book is an exploration of the dispute in Burnt Church, an attempt to articulate what it "really *was* about" for the native and non-native locals involved. When we look beyond fish to the heart of the dispute, we can see that this is a conflict over legitimacy and belonging in a contested place. It is a conflict over whose rules and whose history counts, a conflict over sovereignty and nationalism articulated in the language of conservation. The apparent problem of "resource management" is actually a reflection of the cultural, religious, and political context of Burnt Church. Understanding the dispute in Burnt Church/ Esgenoôpetitj in these terms helps us to see it in context, as one flashpoint on the larger landscape of Canadian colonial conflict.

Re-membering

Acknowledging Canada's colonial history and context is central to developing an understanding of contemporary relationships (or the absence of relationships) between indigenous people and settlers. (See, for example, Alfred 2005, Freeman 2002, Kovach 2009.) As both indigenist and post-colonial theorists point out, the Canadian context is not post-colonial because in Canada the settlers have not left (Alfred 2005, Gandhi 1998). Canada remains a colonial nation, and settlers retain their political power and sense of legitimacy. Nonetheless, the insights and ideas of post-colonial theory, in which people write about experiences of empire, can be helpful in interrogating the nature of colonial experience in Canada. In *Postcolonial Theory: A Critical Introduction*, Leela Gandhi emphasizes the effects of "historical amnesia" on colonial and post-colonial peoples, who want to forget the trauma of their experiences and actions and move forward into a utopian future (1998). She echoes Babha's argument that recovery from the colonial experience requires a historical "re-membering, a putting together of the dismembered past to make sense of the trauma of the present" (9). Re-membering is both historical and psychological, a process of attending to the fractured experiences of the present in order to uncover the forgotten reality of the past. It requires us to "acknowledge the reciprocal behaviour of the two colonial partners," that is, the mutuality of the traumatic relationship and recovery from it (Memmi paraphrased in Gandhi, 11). Across the Americas, indigenist activists and their allies often argue for a process of *decolonization*, a restructuring of power relationships between indigenous people and settlers.

This restructuring must begin from a recognition and remembering of relationships as they are and have been. In *A Fair Country: Telling Truths about Canada*, John Ralston Saul emphasizes the importance and success of relationships between indigenous people and settlers in creating a unique and progressive nation (2008). While Saul's depiction of Canada as a "Métis nation" is appealing, it can also be read as a utopian vision that minimizes the trauma and difficulty of living these relationships in practice. The successful and creative relationships that Saul charts are a part of the story. But these relationships have also been formed by conflict, aggression, and duplicity, as the history of the Metis people themselves demonstrates. Louis Riel, the great early leader of the Metis at Red River, is considered by many Canadians to be a Father of Confederation – and by others to be terrorist or a religious zealot. Elected three times to the Canadian parliament while in exile, he was never allowed to take up his seat. In 1885, the Canadian government executed him for treason because of his leadership of the North-West Rebellion, a movement that attempted to get the Canadian government to recognize the grievances of the Metis. Seeing Canada as a "Metis nation" requires remembering the complicated and difficult nature of relationships past and present.

Relationships between native and non-native groups spark violent conflict and creative resistance across Canada every summer. In Caledonia, the Haudenosaunee/Six Nations occupation of development sites (on territories granted to them as a part of their historic agreement with the Crown) has continued since 2006. Non-native residents and other activists have considered forming a citizen's militia to carry out arrests of and interventions with native occupiers, arguing that it is up to residents to enforce the law since the police refuse to. In Oka, Quebec, in 1990, native protesters erected barricades to block development of a golf course situated on lands under claim, which they believed to be sacred. This precipitated a tense standoff, involving not only native protesters and the provincial police, but also the Canadian army. A provincial police officer was shot and killed, and two other deaths are also attributed to the conflict. In 1995, members of the Stony Point First Nation occupied Ipperwash Provincial Park (Ontario) The park was situated on land expropriated from the First Nation by the Department of National Defence during the Second World War, and the occupation was an attempt to force the government to fulfil its promise to return the land. The unarmed occupiers were confronted by the Ontario Provincial Police, and protester Dudley George was shot and killed by police snipers (Linden 2007). These events are often viewed in isolation from one another, overlooked as anomalous moments in a generally peaceful history.

Such conflicts are often persistently defined in narrow terms, as conflicts over land development which pit native religious values against contemporary economic interests. This sort of thin, single-issue analysis tends to further polarize the situation, as people struggle to express their complex concerns in the simplest of terms. In Oka, Caledonia, and Ipperwash, violence was the eventual result. This is not always the case. In other landmark cases, where people made space for the expression of deeper concerns and the development of long-term relationships and solutions, real insight has been gained. The ongoing conflict between the Cree and the Quebec government over the James Bay hydroelectric development in Quebec persisted for more than thirty years. While the Quebec government was concerned about its ability to build hydroelectric dams, the Cree's fundamental identity as a people was tied to the land in question. For the Cree, the issue was not simply the presence of dams, but their jurisdiction over their own territories. In

2002 the two governments signed La Paix Des Braves, a nation-to-nation agreement which recognized the joint jurisdiction of the Cree and Quebec. This agreement resulted from the eventual recognition by non-aboriginal people that this conflict was not just about dams, but about larger issues of identity and autonomy.

The dispute in Burnt Church/Esgenoôpetitj has often been presented as another isolated political conflict, arising from the peculiar inability of locals to negotiate fisheries agreements with the federal government. The federal government persisted in defining the dispute as a conflict over resource management, and was only willing to discuss questions of fisheries management at the negotiating table.[7] Native activists for the Mi'kmaw fishery were characterized as "lawless." Non-natives felt that the government was powerless or unwilling to enforce the law, and that they had to take the law into their own hands to protect the fishery. These polarized positions have been reciprocally created over centuries of relationship and estrangement. Though they are ostensibly positions about the politics of scientific resource management, these moments are a crystallization of unrelenting conflicts, constructed by the cultural and religious realities of life in the Burnt Churches. The *Marshall* decision precipitated a dispute that reflects the deeper problems of relationship along the Miramichi, and across Canada.

The larger *Marshall* literature provides historical, legal, and political examination of the court's decision, as well as its context. Within a year of the decision, Ken Coates published *The Marshall Decision and Native Rights*, which considers the broader social implications of *Marshall*. Coates argues that the "*Marshall* decision did not cause the deep divisions that have subsequently been revealed between First Nations and other Maritimers" (2000, xviii). He explores the historical and cultural context of *Marshall* and outlines the early events of the dispute in detail (relying largely on media sources). In 2001, Bruce H. Wildsmith discussed his experiences as Donald Marshall Jr's lawyer at trial and in subsequent appeals in the *Windsor Yearbook of Access to Justice*. In 2002, historian William C. Wicken's *Mi'kmaq Treaties on Trial* knit together the story of the peace and friendship treaties with a fascinating account of Wicken's expert testimony for the defence in *Marshall*. While these authors are largely supportive of the Mi'kmaq, there are others who remain concerned about the legal implications of *Marshall*; Aboriginal legal scholars such as Borrows and Henderson, however, seek a fundamental transformation of the legal system and the treatment of Aboriginal peoples within it.

Marshall is considered by Aboriginal legal experts as a critical moment in the ongoing negotiation of the relationship between Canadian and indigenous jurisprudence. For example, in *Canada's Indigenous Constitution* (Borrows 2009), *Treaty Rights in the Constitution of Canada* (Henderson 2007), and *First Nations Jurisprudence and Aboriginal Rights* (Henderson 2006), Borrows and Henderson attempt to interpret and explicate indigenous legal traditions in their own terms, as well as in relation to Canadian law. On the other hand, some non-native constitutional scholars suggest the court's decisions are already far too radical in their implications. Isaac (2001) characterizes the majority's decision in *Marshall* as exceptionally broad and liberal, while Cameron (2009) criticizes *Marshall* as a failure of judicial activism. Cameron suggests that *Marshall* "pits entire communities against one another, ... fundamentally affects the economic interests of those communities, and ... reorients the legal landscape" (10), as though the court's decision and the violent dispute that followed were without legal precedent or historical context.

This book offers a new way of looking at the dispute in Burnt Church; it is an exploration of the implicit values, concerns, and beliefs of local people in the First Nation and in the English (settler) village. Ken Coates's observation that there are deep and historic divisions between First Nations people and (other) Canadians is demonstrated by a close and careful examination of the experiences and world views of those who lived with the conflict in their front yards. Understanding the lived experience of those involved in the fishing dispute – and their beliefs, motivations, and rationalizations – is the priority here.

Place

To local people, both Mi'kmaw and English, the dispute was impossible to understand unless one understood the landscapes and communities where it had occurred, as I explore in chapter 3. If one did not understand the place, one could not understand the conflict. Because place was so important within these communities, it is to this literature that I turn – to contextualize and interpret the stories and experiences of the dispute.

Sense of place, the belonging of people to home/territory/landscape, is something so elemental that it often becomes apparent only when these attachments are jeopardized (Basso 1996, Casey 1993, Malpas 1999, Stefanovic 2000). Much recent exploration of place has been prompted by concern with the increasing uniformity of modern lives and landscapes, where attachments to the particular are homogenized (Frodeman 2005). In "The Geography of Nowhere," for example, Kunstler explores the erasure of difference in our relationships to landscape and place (1993). Concern with dislocation, homelessness, and displacement are key features of the interdisciplinary conversation on place. This is illustrated by Mugerauer's *Interpretations on Behalf of Place: Environmental Displacements and Alternative Responses* (1994), a philosophical and interdisciplinary exploration of place and dwelling, and in Casey's influential *Getting Back into Place* (1993), in which experiences of displacement become important in illustrating the meaning of place and dwelling. Study of place is an effort to understand and address the particularity of the human experience – the construction of identity and "sense of place" through the inescapable fact of our location in particular landscapes.

Scholarly attention to the concept of place goes further than concern with displacement. The study of place – or "topical approaches to knowledge," as Frodeman has characterized it (2005, 1409–10) – is also part of the larger critique of the nature of knowledge as it is produced in the isolated academic disciplines. As an alternative to the reductive, analytic approach so often taken in academic knowledge production, place-based studies also seek "to retain a sense of the whole, seeking to understand the relation between and across the disciplines in a particular place ... One can view a topical approach as stripping the pretensions from types of knowledge that claim to escape the skein of interpretation" (Frodeman 2005, 1410). In Burnt Church/Esgenoôpetitj, place is clearly important. Displacement is a critical issue, and the many concerns expressed by local residents in the dispute cut across the academic disciplines. Thinking in terms of place deepens our understanding of the community and the dispute. One hopes an interdisciplinary approach might, as Frodeman suggests, increase this study's "relevance to people's lives" (1411).

Within the contemporary academy, scholars concerned with place come from many disciplines, including philosophy (e.g., Casey 1993, Chawla 1994, Frodeman 2005, Malpas 1999, 2006, Mugerauer 2008, Stefanovic 2000), anthropology (e.g., Basso 1996, Hornborg in Roepstorff et al. 2003, Ingold in Roepstorff et al. 2003), and geography (e.g., Kunstler 1993, Porteous and Smith 2001, Tuan 1974). In examining the recent explosion of academic interest in place, Relph reminds us that "Heidegger's writings are crucial" in the development of these discussions, and in understanding place in its phenomenological complexity

(2008, 2). Heidegger's focus in *Being and Time* on Dasein (there-being or being-there) is a reminder that the question of the meaning of Being cannot be understood outside of lived experience or apart from the world in which Dasein exists (1962). The integral belonging of human beings to being-there, in-the-world, has encouraged others to explore the phenomenon of place as essentially ontological. "There is no being without being-in-place" (Casey 1993, 313). As Malpas points out, engaging Heidegger, "place is integral to the very structure and possibility of experience" (1999, 32). Relph suggests that, given the "big bang" of interest in place, the best way to make sense of things is to "return to place as a phenomenon of experience, and seek clarity there" (2008, 2).

Place is a way of understanding the web of interrelationships between humans and landscape that shapes both humans and the landscape through time. "Inasmuch as *we are*, we are *in-the-world*, which means that we are always implaced ... To exist is to exist *somewhere*, in some place" (Stefanovic 2000, 103). In this context, place is not simply something that is socially constructed, and not simply spatial or geographic (Malpas 1999, 28–30). The existence of the social *requires* place (36), and just as specific places shape the social lives of the communities within them, these communities shape the places in which they find themselves. Drawing upon the writings of Heidegger and Bachelard, Ingrid Stefanovic argues that the nature of dwelling, of human beings as implaced, is that in which "the reverberation of the hidden and the revealed is fundamental ... [and] resonate[s] with the interplay of presence and absence that defines human existence" (2000, 107). Place is a social and geographic notion that explains how humans relate to the world in which they find themselves; it is also an ontological notion that encompasses our ongoing negotiations between possibility and experience, the space in which our relationship with the existential and originative is articulated – and presumed. There are perhaps at least as many ways that people understand this ontological relationship as there are political systems, languages, or religions and cultures.

In Burnt Church, two communities live in opposition and, at the same time, in relationship, bound by their differences on areas of common concern. Common location – rural, northern, implaced – creates some similar concerns and values for both groups. While these communities find themselves on different sides of many issues, their concerns and priorities are similar: governance, resource depletion, livelihood. The experience of colonization reinforced difference and separation between these two communities, even as it bound them more tightly to one another. Memmi has emphasized the importance of understanding this relationship of mutual desire and hatred between the colonizer and the colonized (1965); Leela Gandhi suggests that explorations of these relationships "are most successful when they are able to illuminate the contiguities and intimacies which underscore the stark violence and counter-violence of the colonial condition" (1998, 11). In this context, investigating place is also a political act, one that engages the colonial dimensions of Canadian life in a particular place as historical and contemporary reality.

When reflecting upon place, it seems tempting to suggest that what we must do is maximize rootedness, reclaim and strengthen the ties of specific people to specific places. But if rootedness means that place must be preserved as a static historical form, then it comes at the expense of difference, as an exclusive practice (Stefanovic 2000, 115). This call to attend to place is not a call for a return to rootedness, to a romantic time, or a mythic age. As Mick Smith has shown, "there is no a priori reason why even extreme modernist narratives do not deserve attention as examples of a world speaking through people" (Smith 2001, 8). Place is not an answer or a principle, it is a condition of being. All world views are bound in place. Developing a critical ethics requires attending to place, requires an articulation of how constellations of factors bind together values, practices, and landscape, and requires a critical evaluation of what this means for those in place, both human and non-human. The notion of place is not one that calls us to a specific perfect relationship to space (or to time) as a solution to our conflicts. Attending to place means attending to the narratives of people and landscapes "where they are at," seeking to understand the explicit and implicit values being negotiated.

Place helps us to attend to the social and environmental relationships that *are there* in communities, in their myriad differences. It requires us to take seriously people's moral positions in and of themselves

> by opening a space for communication of values, and for illumining implicit paradigms that drive a community's very sense of place. Ethics becomes a *dialogical* challenge, rather than a theoretical challenge, to discern a moral order that implicitly instructs a society through its culture, its historical tradition, and the geographical place within which it is situated. (Stefanovic 2000, 129)

According to Stefanovic, a place-based ethic is attentive and open to the narratives of people in their own locations. These narratives will be contradictory and contested, but will point us some way towards, at least, understanding what is critical in people's conceptions of the good. The challenge of understanding place as the ontological grounds within which people develop their notions of the good becomes historical, philosophical, and anthropological. Attending to place requires enquiry into history and landscape, as has been shown. It also requires thoughtful, critical analysis of narratives for converging values, needs, images, and priorities – and for occasions when such needs and values remain in opposition. Dialogue based upon these common and divergent concerns begins to open the way to some understanding of the human needs and values at the heart of the dispute in Burnt Church.

Qualitative Research: Phenomenology, Ethnography, and Lived Religion

This project is *qualitative* in that it seeks to understand experiences and world views in their depth and complexity rather than seeking breadth and repeatability. It integrates phenomenological and ethnographic inquiry, theoretically and methodologically. Broadly speaking, qualitative research contributes to interdisciplinary inquiry through its attempt to understand the richness of lived experience, and to thereby signal the possibility of new, broader directions of enquiry. The point of this book is to develop a critical understanding of the roots of the events in Burnt Church by focusing on peoples' stories of their own experiences and the implicit values which these bring to light. Nurturing a deeper understanding of what has happened in this contested place, in the two Burnt Churches, offers questions and insights that can then be tested more extensively in subsequent studies.

Developing an understanding of the nature of experience requires attending to fundamental philosophical questions such as "Who are we?" (the ontological) and "How do we know what we know?" (the epistemological). Phenomenologists argue that answers to such questions begin not with theory but with phenomena – "that which appears" before theoretical speculation. This is a philosophical movement away from abstract theorizing and "to the things themselves," as Husserl famously wrote. Unlike postmodern theory, phenomenology accepts that some essential structures of experience define our ways of being in the world. Like postmodern theory, phenomenology is socially constructivist, concerned with revealing "the beliefs that shape our lives and what we take as 'truth' and knowledge" (Kovach 2009, 111). We can get to "the world itself," but only through our socially constructed experiences. Environmental philosopher and phenomenologist Ingrid Leman Stefanovic describes it as follows:

> Instead of investigating exclusively subjective experiences or objective, value-free facts in isolation from one another, phenomenology examines the relation between human beings and their world, before philosophers engage in any theoretical abstractions that divide or separate their lived experiences from the world within which these experiences find their meaning and their ground. (2000, xvii)

She suggests that phenomenology is an approach that helps us to navigate between the extremes of naive universalism, on the one hand, and postmodern scepticism, on the other. The phenomenological approach sets out a framework of "critical holism" which attempts to recognize that humans live *within* systems of relationship – with one another and with non-human nature. The methodology attempts to resist reductionism (Marietta 2003) and is concerned with understanding things *in their context* rather than as abstracted or separated from it (Brown 2003, King 1999, Marietta 2003, Smith 2000).

Contextual and holistic approaches to the study of culture and religion are sometimes criticized as overly simplistic, impressionistic, and lacking in critical distance. In *Disrobing the Aboriginal Industry: The Deception behind Indigenous Cultural Preservation*, Widdowson and Howard criticize Aboriginal activists and most academics who engage with them for creating an industry of illusion (2008). They argue that the (so-called) Aboriginal Industry takes unverified knowledge (such as traditional knowledge) as the foundation for self-serving policies and bureaucracies, all the while compiling "whole libraries of 'scholarship' that obscure[s] the actual implications of current aboriginal policies" (9). The Aboriginal Industry reinforces its power with a constructed taboo, "the cry of 'racism' that meets any honest analysis of aboriginal problems and circumstances" (10). Widdowson and Howard suggest that those in the Aboriginal Industry are idealistic, emotionally motivated, and uncritical (21), cynical, or self-interested, like "the anthropologists who encourage a backward spiritualism and mythology in which they themselves do not believe, but which keeps native people in a convenient state of passivity" (22). This argument is troubling, not least because its authors employ a strident rhetoric that oversimplifies and demonizes its subjects. Those of us who persist in discussing cultures and religions holistically are either damnable believers (see pp. 24–9) or self-serving and duplicitous anthropologists. Widdowson and Howard have judged religion a damaging and backward manipulation and deemed it impossible to study. This is strikingly similar to Marx and Engels's depiction of religion as "the sigh of the oppressed creature, the heart of a heartless world, ... the opium of the people" (Marx and Engels in Nye 2008, 59). While religions do function as ideologies, religion is more than ideology.

When religion is studied in its broad "social, cultural, and political roles," the internal organizing principles of values and world views can begin to become apparent (Capps 1995, 158). But how and what shall we study? Sparked by the theories of Durkheim and Weber, academic engagement with questions of the purpose and function of religion has included quantitative and qualitative research. Robert Bellah's explication of the role of "civil religion" in the United States emphasized the importance of religious frameworks in public moral principles and social life regardless of the relationship of such frameworks to formal religions or religious traditions (1967, 1991). Clifford Geertz, a cultural anthropologist, suggested that religion must be understood as a cultural system (1966). He argued that anthropologists must not approach religion as another illustration of anthropological theory, but understand it as a constitutive element of societies and cultures, and in its own terms as "a way of approaching the world or as a mode of engaging reality" (Capps 1995, 182). He emphasized the importance of detailed, narrative-based ("thick") description of culture, and of the role of religion as a source of world view and perceptual framework. This understanding of the holistic social nature of religion underpins the anthropology of religion and guides the ethnographic methodology of this project. It is exemplified more recently by Robert Orsi's work on "lived religion" in America, which emphasizes the importance of understanding religion as a part of everyday life rather than as a rarefied category separate from it (2002). In order to understand culture, then, we must attend to religions as a dimension of culture.

In their criticisms of qualitative scholarship engaged with indigenous cultures and religions, Widdowson and Howard are among those who suggest that all such research is impressionistic and unverifiable. To answer this criticism, I turn to Creswell, who confirms that qualitative researchers *can* verify that their work is "believable, accurate and right" (1998, 193). He outlines a wide variety of methods that qualitative researchers can and should use to verify their results. Overall, he suggests that procedures for verification in qualitative studies include

- *prolonged engagement and persistent observation* in the field, building trust with a community and learning about local culture;
- *triangulation*, or the use of multiple sources and methods to corroborate findings;
- *peer review* or *debriefing* of findings;
- *negative case analysis*, refining the working hypothesis as the research progresses, in light of disconfirming evidence;
- *clarifying researcher bias* through ongoing self-reflection and engagement with other researchers and research participants;
- *participant checks*, soliciting participant's views of the findings and interpretations of the project;
- engaging in *rich, thick description* of events, perspectives, and experiences;
- and seeking out *external audits* of the research and its findings. (201–3)

Cresswell's position is that any study should engage with at least two of these verification methods; in this project, I engaged with them all at some level over ten years of research and writing (as I describe later in this chapter). The ethnographic researcher participates from within and outside the world of experience in a particular place, and recognition of the researcher's subjectivity does not impede the accuracy, believability, or usefulness of the project outcomes. Ethnographic and narrative approaches to problems of values and religion are important not because they render some kind of neutrality or objectivity, but precisely because they can illuminate the complex lived experiences of people in their holistic context. The insights of qualitative research are verifiable, and raise important questions that can be generalized to other research contexts.

"Who Are You, and What Do You Want?" Doing Research in Burnt Church/Esgenoôpetitj

Engaging in research with Aboriginal people as a non-Aboriginal person is a challenging and delicate process – within the academy and within Aboriginal communities. Historically, most academic researchers have carried out work in such communities exclusively for their own benefit and advancement, with little regard for the needs and interests of Aboriginal people themselves. "You anthropologists come in and get what you want then leave. We're still here, and never seem to get anything back in return" (in Richer 1988, 414). In some cases, the bodies, places, and sacred knowledge of peoples have been used for research in abusive and harmful ways. In others, the stories and history shared with researchers became unrecognizable by the time they were published. Our academically "credible" sources are sometimes not only inadequate but harmful to those who find themselves their unlucky subjects.

Indigenous collaborators in academic research can offer important insights and challenges to research methods and methodologies. For Maya activist and academic Sam Colop, the problem of academics evading their responsibilities to the communities in which they work is insidious, particularly in indigenous communities. He summarizes the issues as follows:

1 Foreign [i.e., non-Maya] scholars who do not consult with the community where they are going to work about their projects and who rarely present a final report of the study to the community.
2 The existence of a large body of knowledge about the sociocultural history of Maya communities gathered and compiled by foreigners that is not available to them, leaving local communities ignorant of what foreign scholars have said about their language, culture, or community.
3 Some foreigners who hide religious or proselytizing agendas behind their academic status and who interfere in Maya decision making and others who by not opposing these actions seem to approve them.
4 Foreign researchers who seem only concerned with fulfilling their university or institutional requirements or with gathering data for publication and who take the service of the community for granted. (Luis Enrique Sam Colop [1990] in Warren 1998, 82)

In this context, many challenges arise for those in the academy who seek to understand the relationship between indigenous people and settlers. How can one carry out academic research with native peoples in justice and fairness? Is this even possible, especially for non-native researchers?

In Canada, the Report of the Royal Commission on Aboriginal Peoples (1996) lays out some principles for academic research with native people. The RCAP report outlines the minimum of what is required. Aboriginal activists and academics, and non-Aboriginal academics who have worked closely with Aboriginals on projects, continue to argue for the importance of collaboration, a sense of limitation, respect, and guidance in academic research. Some investigate the nature of indigenous/non-indigenous relationships by examining non-indigenous culture and history (e.g., Freeman 2002, Menzies 1994). Many Aboriginal communities have developed their own ethics review processes for researchers who want to work within their communities, as a way of ensuring that the needs and standards of the local community are being upheld (Armitage and Ashini 1998). In Mi'kmaw territory, the Mi'kmaq Grand Council (a traditional form of government) carries out its own review of all research projects in Mi'kmaw communities.[8] Most recently, Kovach argues that one must do "ethics as methodology," approaching research in indigenous communities as relational and reciprocal (2009, 147–9).

Bruner argues in his article "Ethnography as Narrative" that there is no primary, naive understanding of communities that we later explicate or intellectualize; ethnographers all begin with a narrative in their heads that structures their experience in the field (1997). This narrative is then shaped and formed through many tellings over the different stages of research and publication. Such an understanding of ethnographic process emphasizes the importance of contextualizing the perspective of the researcher, as well as that of those participating in the research project; the ethnographer has a location within the field of research, as do the project participants (King 1999, Warren 1998). The phenomenological emphasis on the relations *between* human beings and their world (Stefanovic 2000) does not exempt researchers, but includes them, in their particular role, in relationship to and yet apart from the community with whom they engage in research.

In *Indigenous Methodologies*, Margaret Kovach argues that researchers need to be clear about their purpose, remembering their own motives and questions and sharing them honestly as a part of their research (2009, 113–15). From as far back as my undergraduate studies in philosophy and biology, I have been interested in the ways that people's values and beliefs shaped what they "knew" about nature. Through graduate work in religion and culture, I continued to explore the relationship between religion, culture, and environment in Canada. In reading about the history of the national parks system, I learned that Duncan Campbell Scott, the bureaucrat responsible for the creation of the Indian residential schools system and the negotiation of some of the numbered treaties, participated in the creation of the national parks system on behalf of Indian Affairs. By engaging with Scott's writings, I realized that the creation of national parks was an implicitly religious process; part of what Scott (and others) believed in was the "triumph" of Christianity over indigenous traditional culture. Over the years I began to wonder whether contemporary conflicts between indigenous people and settlers over resource management might not have their roots in religion, at least in part.[9]

In the summer of 2002, the dispute in Burnt Church was drawing to a close. At the time, I was visiting New Brunswick and Nova Scotia and a friend was involved in the Aboriginal Rights Coalition–Atlantic's Observer Project (ARC-A), a church-based coalition that was training activists to work as witnesses in solidarity with the Burnt Church First Nation. I learned that the Mennonite-based Christian Peacemaker Teams (CPT) had also been working in Burnt Church. It seemed that there might be people in Burnt Church who were already talking about the relationship between religion and resource management conflicts, if these alliances were any indication. When I returned to Toronto, I started to explore the possibility of undertaking research in Burnt Church/Esgenoôpetitj. Two years later, I began fieldwork in the two communities.

The phenomenological and ethnographic research for this book occurred over twelve months in 2004–5 and in follow-up visits in 2007 and 2009. My investigation focused on the social, political, and religious dimensions of people's lives in the communities at Burnt Church; the attitudes and understandings of those who involved themselves in the conflict; and the historical and political construction of these communities and of the conflict.[10] As a participant observer in the English and Mi'kmaw communities, I participated in "regular life," using ethnographic methods to develop my understanding of local life and local values. As the year went on, I also carried out in-depth interviews with individuals in both communities, seeking to engage in a phenomenological elucidation of implicit values and experiences. These two aspects of the research informed and complemented one another.

In reflecting upon my role as a researcher, and the myriad challenges involved, I find that returning to Colop's framework of "ethical failures" is particularly helpful.

(1) Lack of consultation and (2) Resulting knowledge unavailable to indigenous communities

While I sought and followed the advice of people in the communities of Burnt Church, including consulting about written work as I produced it, it is clear that the genesis of this project was grounded principally in my own research interests rather than in the needs of the community. Because of this fact, I sought guidance from and collaboration with community members on the development of the project as it progressed. The collaborative approach requires the researcher to be flexible and open throughout the research process. In this project, the principles of collaboration became important guidelines in developing research relationships with all participants. People within both communities at Burnt Church have important knowledge that provided direction and illumination to the research process and to the resulting analysis.

In practice, this project included a number of collaborations.

(1) *A trip to both communities before the commencement of the project to ensure that the project and the researcher would not be unwelcome.* During and after the preliminary visit, which was hosted by members of the Wabanaki Nations Cultural Resource Centre, some members of the native community agreed to advise and guide me and the development of this project. Within the English community, advisory relationships developed more clearly over time.

(2) *Consultation with community advisers about the focus and approach of the project as it developed.* This included gathering feedback on interview questions before beginning the interview process.

(3) *Attending, where possible, every event to which I was invited.* When I was not invited, I sought and followed informal advice about whether my presence would be appropriate at events that were relevant to the aims of my project.

(4) *Offering to every person interviewed the opportunity to see and respond to the specific parts of the research resulting from their interview before it was completed.* I returned to Burnt Church in June 2007 to share some of my work, and hear people's responses, in person. In order to protect the anonymity of participants during the draft stage, people read primarily the quotes and comments attributable directly to them, and/or discussions of community groups they led, along with an outline of the overall work. A few more interested people read entire chapters or sections of chapters, where anonymity could be preserved.

(5) *Sending copies of my findings, conclusions, and recommendations to people and institutions in both communities, for their information and for further feedback.* In the summer of 2009, I visited the Burnt Churches again, and talked with people about my preparations to publish the research as a book.

Some of the most important collaboration within this project was non-formal and relationship based. As I developed friendships with people in both Burnt Churches, they offered much in the way of informal guidance. Some people also gently refused my invitation to participate in the project, which seemed a natural response after so many years of scrutiny, and as I was an outsider. These refusals were both direct and indirect, and I was not always made aware of the reasons for them.

(3) Researchers who conceal religious or proselytizing agendas

As my area of research includes the study of religion, some presume that I bring a religious viewpoint to my work. This is not surprising since, historically, academic work on religion has been done by religious insiders, and most who enter indigenous communities to "talk religion" have a missionizing agenda. I have already described my approach to the study of religions as an enquiry into human phenomena – as something people do. My professional purpose is humanistic. Does this mean that I am uninfluenced by religion? Of course not. Every human being is influenced by religious dimensions of culture, whether or not they identify as a religious person. Perhaps my own experiences as both a religious insider and an outsider are what drive me to try to understand how religion works and what it means at the level of communities and within cultures.

It is important to describe people's religious lives so that they can recognize themselves, and to think critically about the practices and the world views they embody. In this project, I endeavour to present descriptions that are both sympathetic and critical as a way to get at the depth of people's lived experiences. My agenda is not theological (or, as Colop would say, latently religious) but, rather, concerned with the cultural and social dimensions of human values, beliefs, and practices as they relate to this conflict.

(4) Researchers concerned only with their own achievements and who take the community for granted

When I arrived in Burnt Church in the summer of 2004, it was not clear that this project would have a successful outcome, at least in academic terms. I was not certain that the project (and the researcher) would be well received by the communities involved. Telling people about myself and my project (including some of the reflections outlined above) was my ongoing task. I was fortunate to be welcomed by a number of people in each community quite warmly, if somewhat warily. Re-membering Burnt Church is an academic project in this context, though I hope that it might go some small way towards deepening public discourse on the dispute particularly, and on indigenous–settler relationships in Canada more generally. This tension between academic achievement and community needs is one that runs through much academic work.

I entered Burnt Church/Esgenoôpetitj as a newcomer and outsider – as a person who needed to be taught – in at least equal measure with my position as a researcher or professional. In English Burnt Church, some hoped that my work would finally be the way that "the whole story" got out. Others thought that nothing could really make any difference. In Esgenoôpetitj, expectations of me were never that high, nor so openly low. My trustworthiness is still being established with these written words, and rightfully so. People shared stories and experiences with me in order to educate *me*, and because making these particular stories more public also served their own needs or hopes. There were many stories that were *not* shared with me, and stories shared that I was asked specifically NOT to make public. This project may be the beginning of a deeper relationship, in which a future project could be generated collaboratively rather than simply being carried out in a collaborative spirit, as this one was. The success of this project, limited though it may be, is due largely to those people who took me in as teachers, mentors, friends, and neighbours because and in spite of my research agenda.

Finally, recognizing the discourse about the ethics of research within Aboriginal communities to be instructive in all settings, I carried the methodologies discussed above into my fieldwork in non-native Burnt Church. Among Canadians observing the dispute, I have heard the people of English Burnt Church characterized as simply intolerant racists, "rednecks" who took the law into their own hands. But the views and experiences of this community must not be written off so glibly. If we are to understand the events and effects of the dispute in Burnt Church, we must take seriously the views and experiences of all residents, whether or not one agrees with them. Like most Canadians, the people of Burnt Church live in disputed territory – on colonized lands. Much as academics are called upon to attend to the experiences of indigenous people on their own terms, so also must we attend to non-indigenous others.

As my research took shape, it included participant observation, interviews, a research journal, and collection of ephemera, augmented by the personal archives that two community members shared with me. The participant observation phase of the research extended throughout my entire time in Burnt Church. I was actively engaged in the life of both communities, to the extent that I was welcome, in order to understand the culture and context of the Burnt Churches. I participated in public events – such as the installation of a cenotaph in the English village and the visit of Anishnaabe hockey coach Ted Nolan to the First Nation – and I was actively involved on a regular basis with community groups. My participant observation was primarily with religious organizations, since I felt that these were locations in both communities where people typically sought to explore their values, belief systems, and world views. In the end, I had more extensive informal participation in the life of the English community and participated in more organized groups in the First Nation.

In Esgenoôpetitj, the tensions between the elected (band council) leadership of the community and sovereigntist/traditionalists are an ongoing part of political life. In the post-*Marshall* Mi'kmaw fishery, leadership was provided by political sovereigntists and Mi'kmaw cultural traditionalists, whose power in the community had recently risen, when the elected chief lost his majority on council. In 2004, when I was conducting my research, the elected chief had been returned his majority, and sovereigntists and traditionalists had lost their authority in community political processes. While my research included conversations with people from across these divides within the community, those who believed most strongly in the activism of the dispute were most interested in talking with me, and these people were more often critics of the elected chief. In the English village, where the community is much smaller, the strongest political distinction is between those who are Liberal and those who are Conservative; people supporting each of these parties spoke with me at length.

Getting at world views in a complex way means attending to places as well as to people. Fieldwork comes to involve attending to landscapes as another way to approach place, reflecting on how people and landscape inscribe themselves upon one another. The sources of understanding for an academic project, the *data*, expand to include participant observation of place as well as of people and communities. My research journal came to include my experiences not only with people, but of place as well. Reflection on place frames my experience as a researcher "in the midst of things." As a researcher, trying to understand the Burnt Churches meant not only trying to understand the people and their lives, but also trying to understand the place, the layers of relationship and identity that show themselves on the land. Being in a place opens up the unique nature of that place, much like meeting a person you've only heard or read about. These observations of place become one more way to get at an understanding, such as it may be, of the dispute in Burnt Church/Esgenoôpetitj.

Finally, the research for this project included extensive, in-depth interviews with a few, select individuals from each community. These interviews were designed to be open-ended, directed as much as possible by the concerns and experiences of the interviewee. The inter views themselves usually took well over two, and sometimes three, hours. The participant observation and the interviews, taken together, form the "data" upon which this analysis is based.

The anticipated outcomes of this research include the "thick" description of lived experience, articulation of the role of unacknowledged values and beliefs, and the resulting generation of new questions and insights. To this end, I analysed the stories and experiences gathered using phenomenological and narrative approaches. The phenomenological uncovering of taken-for-granted values and assumptions is primarily accomplished through an analysis of common themes and concerns (see, for example, Creswell 1998, Stefanovic 1994). Narrative approaches to contextualizing these results are commonly recognized ethnographic tools that allow one to situate the values of collaborators in the context of their lives and places (Bruner 1997, Cheney 1997, MacIntyre 1997).

One great challenge of projects that seek to take seriously people's stories and experiences in their own terms is that of putting spoken words to paper. All that I heard in Burnt Church was based upon my increasingly meaningful relationships with people and arose from their interest in helping me understand what had happened in their communities. Everything about my relationships, and about these stories, is lived and oral, and tied closely to the place of Burnt Church. There is much more to these relationships and to people's lives in Burnt Church than I have been able to capture in this book. That fact hardly makes the project valueless, however.

This small window into the dispute does more to approach the lives of local people and the heart of the dispute than almost any other document in the public record. But many more windows into this conflict need to be opened – some of which can only be initiated by the people of Burnt Church and Esgenoôpetitj themselves. Other possible avenues require the careful approaches of my fellow academics. This book represents my attempt to share some of what people tried to teach me, with such grace and patience, during my year in Burnt Church. It is a reflection of life in Burnt Church between two and three years after the dispute subsided, in the years 2004 and 2005. It is one attempt to relate the deeper stories of the dispute, by foregrounding the lived experiences of Burnt Church/Esgenoôpetitj residents and exploring how their deep-seated values and beliefs affected their interpretations and experiences of the conflict – and of their place.

2 "Those Relationships Became Countries"

In *We Get Our Living like Milk from the Land*, the authors suggest that history needs to be seen as more than "just the past" (Maracle et al. 1993, 17). History must be explored in terms of relationships, they argue, "because those relationships became Countries" (17). So, too, the dispute in Burnt Church is more than just the past. The dispute tells us something about the Canada that has developed from those early treaty relationships. The dispute is also, itself, a set of relationships – relationships not only of conflict and confrontation but also of alliance, sorrow, solidarity, and estrangement. This chapter outlines the events of the dispute by integrating public stories of the dispute (media reports, political commentaries, government press releases) with those told by people who lived it. The stories I share here are drawn from my fieldwork with the English and Mi'kmaw communities in Burnt Church.[1] Retelling this story of the experiences and relationships of the dispute is a way of resisting the historical amnesia of Canadian life and of building up the context for the theoretical and analytical discussion of the dispute that follows in chapters 3 through 6.

Esgenoôpetitj, the Burnt Church First Nation, is a community of about 1300 people living on reserve lands on the shore of Miramichi Bay. The community has serious housing and employment shortages; in 2000, the unemployment rate was 85 per cent (Dharamsi 2000). There are many children in the community, most of whom attend the community school on the bay. Those adults who have work are usually employed by the band council, often in the Adult Employment or Community Health Centres or in the ongoing struggle to maintain housing and infrastructure. Next to the Burnt Church First Nation, on the same small peninsula on the edge of Miramichi Bay, lies the village

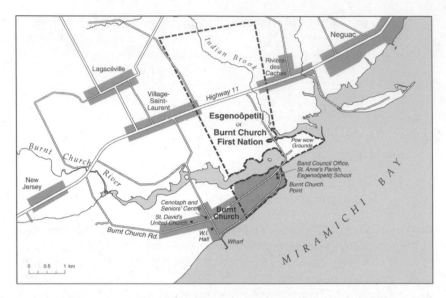

Figure 2. Burnt Church Village and Burnt Church/Esgenoôpetitj First Nation.
(Source: The GIS and Cartography Office, Department of Geography, University
of Toronto)

of Burnt Church. There are about eighty-five residents of this community; they have a church, a community hall, and a small local credit
union, and they have converted the old schoolhouse into a seniors' hall.
People make their living fishing and working in the woods. A few work
in the paper mill in Miramichi and at the local school board; many
joined the Canadian Forces or left to find work in other cities across
Canada, at least for a time. There are far fewer children here; young
people often leave for work or school and raise their families
elsewhere.

How Did the Dispute Start?

In public reports, the dispute began with the Supreme Court of Canada's
decision in *Marshall*, which upheld Donald Marshall Jr's right to fish
and gather wildlife as an obligation under the treaties of peace and
friendship signed by the Mi'kmaq and the Crown in 1760 and 1761.[2]
The court saw the recognition of a Mi'kmaq treaty right to fish as

necessary to "uphold the honour and integrity of the Crown" (*R. v. Marshall*, 17 Sept. 1999, 2). With this decision, Mi'kmaw people began to fish lobster and sell their catch commercially, without government licences and outside of the regulated season. In interviews with local people, I began by asking, "How did the dispute begin?" In both the English and the Mi'kmaw communities, people said the "beginning" of the dispute was long before *Marshall*, though they placed its origins in different centuries.

In the English village, some non-native fishermen say that the problems in the lobster fishery began with the 1990 *Sparrow* decision of the Supreme Court. *Sparrow* recognized that indigenous people had an Aboriginal right to fish under the constitution, based on historical Aboriginal practices (*R. v. Sparrow*, 31 May 1990). After the *Sparrow* decision, the Department of Fisheries and Oceans (DFO) permitted the Mi'kmaw to begin a fall fishery for food and ceremonial purposes.

For many years before *Sparrow*, the Mi'kmaq had been almost entirely absent from local fisheries due in part to the prohibitive costs of buying into the Canadian regulatory regime. This exclusion was typical across the Atlantic. In 1994, in her review of the extensive literature on Atlantic fisheries, Notzke found no mention of Aboriginal people or Indian fisheries "with even one single word" (1994, 61). While there were individual Mi'kmaw fishers across Atlantic Canada, perhaps 150, in Nova Scotia the proportion of Mi'kmaq engaged in fishing, hunting, and related activities declined from 6.5 per cent to 1.4 per cent between 1975 and 1980, while the percentage of non-native people so engaged remained relatively constant (Notzke 1994, 61; Wien 1986, 65). Historically, the Mi'kmaq relied on the fisheries as one of their primary sources of food, and fishing was a central part of Mi'kmaw social structures (Upton 1979). Many academics and native activists now recognize the systematic exclusion of the Mi'kmaq from the Atlantic fisheries as a part of the injustice of a colonial government's disregard for treaties (e.g., Coates 2000, Obomsawin 2002, Wicken 2002).

The non-native residents of Burnt Church see the absence of a Mi'kmaw fishery as normal and as the responsibility of the Mi'kmaq themselves for "giving up" their place in the waters. Non-native fishers and their communities believed that the DFO overemphasized the cultural importance of fisheries to the Mi'kmaq when implementing *Sparrow* and therefore allowed them unfair access to the lobster fishery. Nine years before *Marshall*, when the Mi'kmaq began their fall lobster fishery under the terms of *Sparrow*, non-native fishers began to feel that their access to the fisheries was under threat from what they believed to be "special privileges."

People in Esgenoôpetitj argue that the roots of the dispute are far deeper even than *Sparrow*: they lie in the history of community access to resources, in Indian Act politics, and in the colonial history of Canada. Kwegsi (Lloyd Augustine) is the keptin (traditional chief) at Esgenoô-petitj. He was also the co-facilitator of the community's consultation around the fishery and the co-author of the resulting *Draft for EFN* [*Esgenoôpetitj First Nation*] *Fishery Act* and *Draft for EFN Management Plan* (Ward and Augustine 2000a, 2000b). A man of strong sovereigntist opinion, Lloyd argues that the roots of the dispute in Burnt Church are centuries old: "The dispute began in 1725 with the signing of the first treaty.[3] It began with those treaties, and it goes right on down to today. It didn't begin with Marshall, no matter what some people will tell you." For Lloyd, the post-*Marshall* conflict in Burnt Church is a continuation of a historic dispute over the nature of the relationship between Canada and the Mi'kmaq, a relationship framed during the negotiation of the peace and friendship treaties. Lloyd understands the dispute as one between and about the parties to the treaties.

> It's been generations of this kind of thing for our people, the Canadian government takin' away our lands, not respectin' our sovereignty over our own territory, tellin' Canadians that the land is theirs to do with as they please. They don't recognize that the treaties they signed were trea-ties between Nations. These are all Mi'kmaw lands. We've been put on this reserve by a government that is just trying to get rid of us, to wear us down or to kill us off, hopin' we'll keep quiet. These are our waters, not Canadian waters.

Miigam'agan, a former elected band councillor and one of those working to revitalize traditional culture, echoes Lloyd's long and criti-cal view of history, suggesting that the violence of the dispute has its roots in Canadian denial of treaty relationships. In a letter, she writes:

> I really believe this has been a long time coming, the canadian citizens have long held a fear of losing what they have acquired on Indian Lands. And now they felt betrayed by their government and their law ("canadian supreme court") who finally acknowledged the treaties with Mi'kmaq ... Our community has always been aware of *tan wen nenan ag tan teliagu'p*, our inherent right and the history of our people, and the negative impact

since arrival of the europeans.[4] From this experience there is a common fear among our people that if we, as native people, exercised our inherent rights we would be attacked by the canadian governments and its people. This is a common knowledge as continually proven in history. [punctuation as in original]

For the people of Esgenoôpetitj, the roots of the dispute also lie in the immediate past, in a community revitalization movement. About ten years before *Marshall*, a small group of people began to revive traditional teachings and culture in an attempt to improve life on the reserve. This group was a part of the movement to host an annual powwow in Esgenoôpetitj, to establish the Wabanaki Nations Cultural Resource Centre, and to teach children traditional culture.

As a result of this work, the community elected a new band council in 1999 who also held the same vision of a healthy native community ... In 1999 when Donald Marshall's case won at the Supreme Court, our community was also in transition from old regime to a newly elected council. There was a strong sense of hope in the community who was also at the time still recovering from the aftermath of a band election and from the wrath of the old council. (Miigam'agan)

Wilbur Dedam, the long-standing elected chief, did not lose the election entirely, but he did lose his majority on council. The importance of the changing role of the band council in 1999 was echoed by non-traditionalist community members Dalton and Cindy. They recall that earlier court decisions that upheld the rights of the Mi'kmaq to harvest timber had resulted in little new employment on the reserve, and blamed this on agreements the elected chief and council made, giving themselves control over contracts. Since control of the contracts remained with the chief and council, all resulting employment and income remained with them, rather than being distributed throughout the community. To some degree, these earlier court decisions had in fact reinforced and entrenched political and economic disparities in the community, rather than alleviating them.

The *Marshall* Decision

When Donald Marshall Jr won his case at the Supreme Court, the ruling was handed down on 17 September 1999. The Mi'kmaq of

Esgenoôpetitj received the decision in the context of their understanding of the treaties and their relationship with their own elected council. People began fishing almost immediately. According to Dalton and Cindy, people "went to the council and told 'em, 'Don't you mess around with this, we're going out there; we're going to fish. You're not going to do the same thing you did with forestry.' And they just went out and started setting traps and everything." The lobster fishery in Burnt Church grew quickly after *Marshall*. The community was located right on the shores of Miramichi Bay, allowing people to essentially fish from their front yards using whatever small boats they had to hand. The renewed power of the community over the band council meant that people were fishing on their own terms, in a commercial fishery that they believed was validated by the court in *Marshall*.

Mi'kmaw people across the Atlantic region began to fish lobster. The three most active fisheries developed where reserves had easy ocean access, at Yarmouth, Nova Scotia, and in Burnt Church, New Brunswick, or in the case of Indian Brook, where people had water access at St Mary's Bay. The growing Mi'kmaw fishery, unregulated by the federal government, caused concern and upset in non-native communities. By 30 September, the *Canadian Press* characterized the mood in the region as developing "against a backdrop of growing turbulence in the Maritime lobster fishery, and fears of violent confrontation on the water between native and non-native fishers" (Morris 1999). Regional chiefs appealed for patience and argued that the small numbers of native fishers in the water would not have a large impact on the existing commercial fishery. Chief Lawrence Paul, of Millbrook, Nova Scotia, suggested that "the non-Indian fisherman must realize that we have the law of the land behind us now" (Morris 1999).

On 1 October Herb Dhaliwal, the minister of fisheries and oceans, released a statement in which he appealed for calm and emphasized that while the court's ruling upholds Mi'kmaw treaty rights, "it has also made it clear the exercise of the right is subject to regulation by Government" (DFO 1999). In the Moncton *Times & Transcript* the following day, the Maritime Fishermen's Union spokesperson argued that 160,000 kilograms of lobster had already been fished by native fishers: "At the rate at which the removals are going, it's clear that there will not be a sustainable commercial fishery in that area in the springtime" (Porter 1999, C2). Within two weeks of the *Marshall* decision, tensions had seriously escalated between those who saw the court decision as an affirmation of treaty rights and relationships and those who saw a native fishery as a threat to resources and regulatory regimes.

Tensions Ignite: The Sunday Protest

The non-native residents of Burnt Church village decided to make a public statement against the native fishery. They organized a protest to take place Sunday, 3 October, in the afternoon after church, and invited people from other local (non-native) fishing communities to join them. The fishers organized a flotilla of boats from many villages, scheduled to arrive at the wharf in Burnt Church at the same time that other locals marched onto the wharf. On the wharf, this group of about twenty non-native people carried a Canadian flag, along with signs with such messages as "What about our rights?" The non-native protesters did not expect that their demonstration would trigger violent confrontation. It did. While protesters were gathering on the wharf, some of the fishers on the water began to cut native lobster traps,[5] destroying them and other fishing equipment. To this day, the white residents of Burnt Church insist that they didn't cut the traps, fishers from other villages did. "But everything came back on us, because everyone knew we organized it" (Mark).[6]

As word reached the reserve that their equipment was being destroyed, people started to come down to the wharf to investigate.

> When we heard the news we went to the wharf and waited for our folks to come ashore and discovered that everyone's traps were destroyed. At the same time the families of the Burnt Church white fishermen came marching on the wharf to protest against the Indian fishing. They already knew … While the wives were coming to worship, the husbands had already gone out and destroyed all of our equipment. (Miigam'agan)

Native people were also shocked to see that some of the men in the boats had dressed up in Indian costumes, with wigs and fake war paint, waving tomahawks in the air, dancing, and yelling racist taunts.

The tension in the two communities erupted. On the wharf and along the shore road people were yelling and fighting. Fist fights broke out; people threatened one another's lives and properties. Natives yelled, "Go back where you came from!" Non-natives screamed, "No special treatment!" Each side felt that they were being victimized by the other.

Eventually the RCMP escorted the white protestors off the wharf, suggesting that this was for their own protection.

> We [Mi'kmaw people] stayed back and decided to take over the wharf. Because of what we had just witnessed from the RCMP's reaction and the comments against our people, we knew there was not going to be justice for us. We needed to respond to what happened to us, but how? We organized ourselves and had a community meeting at the wharf. The people started to feel a little relief after talking about their experience and knowing we were going to stick together. (Miigam'agan)

The natives occupied the wharf.

Violence spread quickly that night. Groups of non-native fishers stormed three area fish plants accused of processing native-caught lobsters. On the Burnt Church wharf, trucks left by non-native fishers were burnt by native protesters. At the same time, one non-native boat was protected by a couple of Mi'kmaq who drove it away from the wharf for its owner. In an effort to replace their destroyed equipment, native youths attempted to raid the equipment shed of one local fisher, ramming the doors with their truck. In the ensuing altercation, a young native man was seriously beaten with a baseball bat. By dawn on Monday morning, the smouldering wrecks of trucks were being towed from the wharf, people were visiting their family members in hospital, those arrested were calling their lawyers and trying to get bail, and the people of Esgenoôpetitj were looking for a way to replace their destroyed equipment.

In the English community, after the explosion of events on Sunday, people felt tremendous fear. The RCMP emphasized the threat to the white residents, telling some that they were "on a list" of people targeted for retaliation by natives. People felt it had taken the RCMP a very long time to respond to their calls for help, and that it was up to them to protect themselves. In almost every home, people sat up at night with their rifles. As the week went on, many had RCMP cars parked in their driveways overnight; some people left their homes for their own safety.

> When that first happened, my husband would sit in that chair, with a rifle across his knee all night. All night long. One night he said to me, "You can't go to bed. If something happened, I can't get you out. I can't protect you, if you're in bed with just your pajamas on." So, for two nights I sat up

with all my clothes on. Finally I said, "This is dumb. If they're gonna kill me, kill me! Get it over with! Why should I suffer like this?" So I said, "I'm going to bed. If you wanna sit there with that rifle, you sit there with that rifle, but I've had enough! If anybody comes in, I'll hear them coming, I'll be ready!" And I did, I kept a .22 in the bedroom, with shells in it. If anybody hadda come in, I wasn't – I wouldn't use a big gun that would kill somebody, but I sure would hurt them as much as I could. Anyway, I went to bed that night, and that night about two o'clock in the morning, the RCMP was sitting in the yard here, and they came up and knocked on the door … And they said, "We just came to tell you that we're going to sit here in your driveway all night long, and for you to go to bed, get some sleep. (Mary)[7]

In the tense days after the Sunday protest, no one was confident of their safety. The violence in Burnt Church was the lead story in the national news; the media descended.

While Mi'kmaw people continued to fish across the region, residents of Burnt Church and Esgenoôpetitj continued to wrestle with the strong emotions ignited by the protest. English villagers were unsettled by the idea that they should "go back where they came from"; Burnt Church was their only home. Mi'kmaq community members were troubled by the suggestion that they were receiving "special treatment," as they understood the fishery as their Nation's right under the treaties. Media attention did not focus on these concerns, but on the responsibility of the court and the federal government for the "lawlessness" in Burnt Church. Photos and videos of the wharf occupation, such as the one on the cover of this book, were broadcast across the country and described in papers: "Mi'kmaq erected teepees and flew Mi'kmaq, Mohawk and Burnt Church band flags. Several natives dressed in military camouflage gear stood by, saying they would block any further non-native attempts to wreck traps"[8] (Poitras 1999b). These images, along with the ongoing threat of violent retaliation in both communities, drew the attention of columnists and commentators. In the New Brunswick *Telegraph-Journal*[9] on Wednesday, Dalton Camp blamed the double-speak of the federal government for the crisis; lawlessness broke out because of the failure of the DFO to act, and of any government official to speak up and say that the native fishermen were legally within their rights (1999). On Thursday in that same paper, Chantal Hébert blamed the Trudeau government, the constitution, and the Supreme Court for the conflict, arguing that "over almost two decades of interpreting the 1982 constitution, the country's top court has regularly crafted out rights where many thought none had existed" (1999).

In the midst of this debate, the federal government urged the Atlantic Policy Congress of First Nation Chiefs (APC) to impose a moratorium on post-*Marshall* fishing, so that some sort of agreed-upon regulation could be reached. Herb Dhaliwal met with regional fishermen's groups, and then with the APC in Halifax. He argued that "a treaty right is a regulated right – [that] the Federal government and I, as Fisheries Minister, can regulate that right," and that he could (and would) choose to impose a solution if that became necessary. Native leaders heard this as an "ultimatum" (Poitras 1999a, A1, A2).

Many in Burnt Church believed that the chiefs of the APC were not consulting appropriately with their people, and that they were being manipulated by the federal government.

> In witnessing this development between the native leaders and canada, the people in Esgenoôpetitj advised the newly elected council to leave the APC meeting and to come and join the community's protest and exercise our right to fish. Although the majority of the chiefs [APC] were persuaded by Canada's/DFO promises, the people in Esgenoôpetitj boycotted these talks. We knew that most of these chiefs were not consulting with their communities nor have they any voice with the Feds. (Miigam'agan)

In the end, the APC agreed to ask the communities involved in fishing to voluntarily begin a thirty-day shutdown. Most Mi'kmaw communities agreed – though most of these were land-locked communities that had not entered the fishery in large numbers. The Burnt Church First Nation did not agree to the moratorium. Reporting in the *Montreal Gazette*, Rick Mofina suggested that this refusal was because native fishers "believe it is their right and duty to support their families in accordance with last month's Supreme Court ruling upholding their ancient right" (1999). While Mi'kmaq in Burnt Church, Indian Brook, and Yarmouth continued to fish, all others in the region pulled their traps out.

Fisheries minister Dhaliwal attempted in mid-October to impose regulations on the Mi'kmaw fisheries in Yarmouth, Indian Brook, and Burnt Church. While the fishermen's unions and other non-native groups welcomed the trap limits as evidence that the government was doing its job and urged them to enforce these limits strongly and quickly, the people in Burnt Church/Esgenoôpetitj received these regulations as a further threat to their rights. They refused to recognize that the federal government had any authority over their fishery, arguing that they were fishing to support their families as outlined in the treaties and affirmed by *Marshall*.

At the Indian Brook fishery, non-native fishermen brought their boats into the harbour in a protest that was also a blockade of native fishers. In Yarmouth, non-natives hauled up and destroyed native traps. In Burnt Church, the government took concrete enforcement action against native fishers. Overnight on Thursday, 21 October, and well into the day on Friday, a fleet of government boats, including Coast Guard cutters and DFO enforcement vehicles, worked in the bay destroying the traps of the native fishery. A DFO spokesman said that to maintain "an orderly and regulated fishery ... We removed what was in excess [i.e., excess traps]. We are satisfied that there are no more than 600 in the water now" (Canadian Press 1999). The anger and tension in the Burnt Church First Nation only rose with this decision of the government, whom they saw as acting against the treaties yet again, denying historic rights and defying the rule of law as laid out by their own courts.

In Yarmouth, remarkably, the situation shifted away from violence. The fishermen's unions had thrown up their hands at the government's management of the situation, saying publicly that they would do better making a deal with the native fishers themselves. Behind the scenes, non-native fishers had approached Chief Deborah Robinson, proposing that native fishers should join them and that they would fish together in the commercial season, which opened in the late fall. The people of Acadia First Nation accepted and entered into an unwritten trial agreement with their non-native neighbours. The flotilla in Yarmouth harbour broke up, and a tenuous and positive new attempt at local cooperation was begun. At the end of October, native fishers in Burnt Church and in Indian Brook began to remove their traps. It was the end of their season. But the battle over *Marshall* continued.

Surprise! *Marshall II*

In mid-October, as Dhaliwal prepared to crack down on the fishery in Esgenoôpetitj, the West Nova Fishermen's Coalition (representing fishers in one region of Nova Scotia) filed a motion at the Supreme Court for a rehearing and stay of *Marshall*. As an intervener in the original case, the coalition felt that the decision of the court was having a drastic and unforeseen impact upon the lobster fishery, and that the court needed to reconsider its decision. Until that time, the coalition argued, the court should issue a stay of their decision in *Marshall* in order to protect the fishery.

On 17 November, after all the traps had been pulled and the immediate tensions were lessening, the Supreme Court of Canada denied the West Nova Fishermen's Association's motion for a rehearing and a stay. Yet, with that denial, the court made an unprecedented move. Before *Marshall*, the court had never provided reasons for refusing to hear a case; they provide opinions only when they issue rulings. But in November 1999, when they refused to rehear *Marshall*, the justices simultaneously issued a thiry-nine-page "clarification" of this dismissal and of their earlier ruling.[10] In part, the court said that the "acquittal [of Donald Marshall Jr] cannot be generalized to a declaration that licencing restrictions or closed seasons can never be imposed as part of the government's regulation of the Mi'kmaq limited commercial 'right to fish.'" The clarification continued:

> The federal and provincial governments have the authority within their respective legislative fields to regulate the exercise of a treaty right where justified on conservation or other grounds ... The paramount regulatory objective is conservation and responsibility for it is placed squarely on the minister responsible and not on the aboriginal or non-aboriginal users of the resource. The regulatory authority extends to other compelling and substantial public objectives which may include economic and regional fairness, and recognition of the historical reliance upon, and participation in, the fishery by non-aboriginal groups. Aboriginal people are entitled to be consulted about limitations on the exercise of treaty and aboriginal rights. The Minister has available for regulatory purposes the full range of resource management tools and techniques, provided their use to limit the exercise of a treaty right can be justified on conservation or other grounds. (*R. v. Marshall*, 17 Nov. 1999, 3–4)

At the time, the CBC characterized the clarification as one that "limits *Marshall*," as did the Government of New Brunswick and other parties. As such, New Brunswick welcomed the decision, and the CBC suggested that the position of the governments as they entered into further negotiation was "one of strength" (CBC 1999a).

Native leaders were much less positive about the implications of this decision. Sheldon Cardinal, a specialist in treaty law at St Thomas University, was surprised at the detailed "clarification" issued by the court. "It is very disappointing to have this happen, to have the courts have an intervenor who had no business bringing this sort of issue to court and have them say no, but we're going to rule on it anyway and somehow limit the treaty right even more" (CBC 1999a). In Nova Scotia, Chief Lawrence Paul said

I'm flabbergasted on it. Why would the Supreme Court of Canada cave in to vigilante and mob rule, and people taking the law into their own hands, destroying public property? What message are they sending out here to the Canadian people?
  … We are not going to get the justice we so desperately strive for unless our treaties are interpreted by an international court that's neutral. We'll never achieve that. We may get piece-meal justice, but we will never get the justice we really think we should have by the virtue of our treaties. (CBC 1999b)

People in Esgenoôpetitj were also shocked by the court's move, which they saw as evidence of the undue influence of the government and the corruption of the Canadian state. "We're considered lawbreakers, that's what we were, and yet we were following the law. The Supreme Court said it was okay. So what they [the government] did was went back to the Supreme Court. Had it changed – they amended it" (Dalton and Cindy).
    The Supreme Court's clarification came after fishers in Burnt Church and Indian Brook had removed their traps, and those in Acadia and Yarmouth had agreed to fish together. Winter was well on its way. For a few months, the waters were quiet. The activity was in government hearing rooms, as the federal government's Fisheries Committee tried to parse out the meaning of *Marshall* and its "clarification," and at the negotiating table, where native leaders and fisheries negotiators were trying to come to an agreement before spring. In Esgenoôpetitj, community members undertook a community consultation process in an effort to provide the elected council with clarity about the collective wishes of the community.

The Continuing Dispute – 2000/2001

In the year 2000, the dynamics of the dispute shifted. While the violence continued, and perhaps even escalated, the roles of the players changed. Confrontations were no longer primarily between local Mi'kmaw and English residents of Burnt Church, but between the Mi'kmaq and the

Canadian government. The government, unwilling to discuss the issues the Mi'kmaq believed relevant (i.e., treaty rights), became increasingly frustrated with the ongoing fishery, and its inability to resolve questions of fishery management.

As commercial fishers looked towards the spring commercial fishery in early 2000, the federal government and the First Nation were at the negotiating table. The negotiations were troubled. The people of Esgenoôpetitj were interested in negotiating in a context that recognized their rights and sovereignty; the DFO, negotiating on behalf of the federal government, was willing to talk about fisheries management. On 15 March, six weeks before the start of the spring commercial season, the negotiators for the Burnt Church First Nation walked away from the negotiating table. The community had decided to develop its own plan to manage the fishery, in Mi'kmaw terms. By the end of April, James Ward and Lloyd Augustine had authored the *Draft for EFN [Esgenoôpetitj First Nation] Fishery Act* and the related draft *Management Plan*, under which the community would issue its own tags and regulate its own fishery (2000a, 2000b).[11]

Over the next two years, all attempts at negotiation between the government and the First Nation were equally troubled. In each case, the negotiator sent by the government was unable or unwilling to address issues of rights and sovereignty, which were a priority for the people of Esgenoôpetitj. Government negotiators continued to talk mainly about fish and money. One clear example of this, for people in the native community, was the arrival of mediator Bob Rae in the fall of 2000. Rae came into the community as a mediator chosen by both the government and the First Nation. People were hopeful that he would be able to make progress, but they found that he was prevented by the terms of his appointment from addressing their fundamental concerns. They came to believe that he was also unwilling to engage with them and their issues. Writing for the CBC, Fenton Somerville describes his own reaction at Rae's first community meeting as moderator:

> To tell you the truth, I was not impressed. He sat there and scratched his head a lot and really did not appear to say much, while the native leaders stated their case. At first, I thought he was listening as a mediator should, but it was the manner in which he listened that made me think. He looked weary and, to put it bluntly, bored. I thought at first that it might be jet lag or the boat tour he took of the bay earlier that afternoon, but no. I sensed he was forced into this situation against his better judgement. It seemed

like his heart was not in it. It's just not what I pictured a mediation process to be. (Somerville 2000)

After only nine days in the community, Rae said there was nothing he could accomplish, and he left. From the perspective of the Mi'kmaq, it seemed that there was no one in government whose "heart was in it," who was willing to seek a resolution of all issues.

The second major challenge to negotiation was the ongoing government raids of the native fishery and the escalating violence on the waters of Miramichi Bay. In the spring, as native fishers began to set traps with EFN tags, the DFO began to raid the community's fishery, hauling out EFN tagged traps. Native fishers persisted, in an effort to assert their rights to fish under the management plan according to their understanding of *Marshall*. A coalition of regional and national environmental and social-justice groups issued a public statement exhorting the federal government to recognize the legitimacy of the native tags. Instead, the government escalated its enforcement, removing traps in early morning hours before daylight.

Native Rangers (fisheries enforcement officers) and Warriors met the government boats in boats of their own, in attempts to protect their traps. Leo Bartiboguc, who led the Esgenoôpetitj Rangers, describes early efforts to stop the raids:

> The Department of Fisheries started coming around [taking our traps], and it started happening every day. They keep taking and we keep putting, taking and putting. It got to the point where we're not only intercepting them, but we're trying to make it difficult for them to be taking our traps. So we used different tactics, like putting little letters and toys in Ziploc bags in the traps, just to try and let them know that they are taking food away from our tables and a livelihood from our children. But it got to a point where they got more and more intense ... The women stopped going in the waters because now it was getting more dangerous because they were coming with a different kind of aggression, and it intensified every time they came.

People in Esgenoôpetitj began to feel that the greatest threat they experienced was not from their neighbours or from commercial fishers, as it had been in the previous year. Now, they were most frightened of the government, and of the armed presence of the DFO, the RCMP, and the Coast Guard.

When the fall fishery began in August, both sides issued threatening statements:

> The Department of Fisheries has said it has as many as 600 officers on standby, ready to enforce the law if anyone tries fishing without licences and federally issued tags. The band, meanwhile, has said its own reinforcements are only a phone call away, including warriors who are not easily intimidated.
>
> "We have a lot of young men who are more than willing to get together and protect our traps," [James] Ward told CBC News on Thursday. (CBC 2000)

By 14 August the confrontations between the DFO and native fishers had escalated. SWAT teams and police dressed in riot gear patrolled the waters and the reserve boundaries. Violent altercations began to break out between these officers and those appointed by the Mi'kmaq. Community members kept careful tabs on encounters Rangers and Warriors had with the government out on the waters.

> When my sister's husband was taken by the DFO, they maced him. And they couldn't knock him down – they were trying to knock him down so that they could get him on the boat. James Ward [a leader of the Warriors] was the number one enemy. They were saying over the radio "We caught James Ward! We caught James Ward!" And my sister's husband could hear them. It was him they were battling with, not James Ward! They were hitting him with billy clubs and whatever they had, and he said, "I just reached back, and the first thing I could grab was a 2x4." He started swinging away at them. And he got a couple of them real good I guess. When they took him to court, they charged him for assaulting nine officers. They were the ones that were beating him up! And they charged him with assault![12] (Barb)

Stories of people's experiences on the waters galvanized the community. By mid-August, people had erected barricades at the edges of the community, blockades marked by bonfires and patrolled by Warriors who had come to Burnt Church from across Indian Country.

DFO officers continued to remove native traps and arrest native fishers. Native fishers told reporters that DFO officers confronting them on the waters had pointed guns at them, which the DFO denied. The public credibility of the DFO was compromised, however, as the violence

Figure 3. After a DFO boat sank a small dory carrying four Mi'kmaw fishers, other Mi'kmaw raced to rescue their community members from the waters of Miramichi Bay. Some confronted these riot-gear-clad RCMP officers and threw rocks at them. (Tuesday, 29 Aug. 2000. CP Photo/Jacques Boissinot)

escalated. On 29 August, a government boat chased a native dory across the waters of Miramichi Bay directly in front of the community. This had happened before – as Rangers and Warriors attempted to confront government officials and prevent the seizure of their traps, they would often be chased off. The shore in front of the community was where local people gathered, and where the TV networks had set up their cameras in position to capture the ongoing conflict. This day, with cameras rolling, the larger DFO boat rammed and sank the native dory. As people scrambled to rescue the capsized fishers, the DFO officers on the boat drove their boat over the dory and the men in the water underneath it. These events were broadcast internationally, and drew the attention of many to the events in Burnt Church.

The violence on the waters continued for the rest of the fall fishing season. On land, the community was under constant surveillance from the RCMP. Listening posts were set up to monitor telephone calls; helicopters flew daily monitoring patterns over the community; people were followed by marked cars when they left the reserve and "visited" by RCMP officers in their homes. In the English village, RCMP officers sometimes explained surveillance methods to people in detail, in an apparent attempt at reassurance. "The RCMP had things on hydro poles they could pick up people's conversations and things, walking along ... They were round the community, but you didn't know where they were ... They had planes in the air that could pick up movement and heat and voices" (Mary). Some of these methods, like the helicopters and the phone taps, were more intrusive, and therefore obvious to those being watched. In the First Nation this was especially difficult for parents whose children were scared by the persistent presence of the helicopters over their community.

Stories of these days are told in Esgenoôpetitj to illustrate the effects of life under surveillance and the constant threat of violence. Here, I share three of them in an effort to bring some of this reality to life. In the first, Leo talks about his work on the waters with the Rangers, after his brother was beaten by the police. He was ready to fire at the DFO in retaliation, but his sister held him back:

> My brother got beat up on the waters, and it became personal. He's been hurt, and somebody whispers in your ear he's in the hospital ... And you know they're still in the water, the very same people that just put a beating on him. So you look at this group of boys that's looking at you and you say, "All right, let's go finish it." And when I said "Let's go finish it," that didn't mean let's go throw rocks over there. It meant let's go finish it. And so they were all jumping with joy, these boys that were ready to go ... They wanted that word ... All we gotta do is just let 'er rip, just start spraying, shooting. That'll be it. All the army would have come in, and everything. That didn't dawn on me because my anger was taking over. But I remember that one word my sister told me was "What kind of a leader will you be to our people when they're dead, or they're in jail? What can you do for the people then?" So ... those are the things that I had in my mind. When we were going down I changed my mind. (Leo)

Lloyd Augustine shares another story about the tremendous risks that people took in those days:

> Everything leads to the fact that we are beaten to the point that there's no self-esteem ... Yet there are others who have connected back to their spirituality, they're starting to stand up and say, "No. I will stand for my

people. Even if it means I have to die for my people." One of the bravest things I heard, and at the same time the most foolish thing was when one of the guys, when they were in the waters, went face to face with these [DFO] guys, who were shouting at them. [He said to the DFO,] "I'm ready to die this morning, are you?" And they just drove off. And our guys just stood there.

One of the women in the community, a band council employee, talks about her experience of this violence, and how she came to see the Canadian government in a new light.

It's almost like, maybe somebody's gonna realize pretty soon that they made a mistake and they'll apologize. But no, nobody came, nobody apologized. The government wouldn't talk to us. We had sent letters to the Prime Minister asking him to do something. No response. We even contacted our union officials, and they contacted the Prime Minister's Office … but no response. I wanna know what was preventing him from getting involved. The way I was looking at it was, okay, they don't see us as Canadians. They're looking at us as people that they have to put up with, because they moved to this country and they don't know how to deal with us. (Barb)

The ongoing violence and threat of violence had begun to change the face of the community, as people felt its destabilizing effects.

The sense of isolation created in the native community by the conflict was deep. The community invited other activists to join them, in solidarity with the fishery. Native and non-native activists arrived from across Canada. Some of the first people to arrive in the community were allies from neighbouring Mi'kmaw communities. Miigam'agan specifically highlighted the importance of this support:

We began contacting other native communities and non-native allies from different organizations and churches. The first to respond was the Listiguj [Restigouche, Quebec] Mi'kmaw community, Chief Metallic and his council and they also brought their Mi'kmaw Rangers. They told the community they responded because of a protocol between Mi'kmaq to unite when one of our communities are in distress. The spirit in the community was uplifted and validated by their presence … Once the media aired the news of what was happening to Esgenoôpetitj and the Listiguj Rangers by the attacks from the DFO, more supporters came. (Miigam'agan)

People came from Elsipogtog (Big Cove, New Brunswick) with boats and equipment. Under the leadership of James Ward and other local men, Warriors from across Indian Country arrived to support and protect the community in its confrontation with the government. Warriors are young native men who are prepared to take up arms in a militant stance to protect indigenous communities. Many of the Warriors who worked in Burnt Church had military training. Their presence in the community was received with mixed feelings. While many were profoundly encouraged by the Warrior's arrival, others felt that this positive presence was tempered by the ongoing challenge of controlling militant factions within the Warrior community. As one man said, "It kind of made me nervous to have them here, and then it was kind of safe to have them here" (Dalton). At the invitation of the community, the Aboriginal Rights Coalition-Atlantic (ARC-A) and the Christian Peacemaker Teams (CPT), non-native religious coalitions, sent in groups of activists to monitor the activities of the government in solidarity with the Mi'kmaq. These activists would travel in the Rangers' boats with them and sit on the shores, keeping careful records of the events of the fishery and the actions of the government.[13]

With great pride, the community also received visits from Ovide Mercredi, the former chief of the Assembly of First Nations. Many people spoke about the great importance of his presence during the dispute, both to them personally and for Esgenoôpetitj as a whole. Miigam'agan described her nervousness about his arrival, and her joy at the way his presence brought the community together:

> The large presence of Ovide Mercredi was very strong medicine for the Elders and the people of Esgenoôpetitj ... There were so many people who came to greet him and then all the Elders came, even ones who were in their homes for so many years. They all came out. There were so many people I was so inspired by my community. At first I wanted to pretend to Ovide that this was like a natural happening, but it was amazing, the response he got, and how respectful he was to the elders and to the women in the community.
>
> ... The Elders brought gifts to Ovide. Nobody organized this presentation and it happened. I was overwhelmed and humbled greatly by this act. I thought, my God, I can't believe our whole community is here. And this is what we don't see about our community. So for me, my whole experience was like I got to see my community in its best light. In what we're capable of being and who we are.

The presence of all these people on the reserve served as a reminder for residents that there were people outside of Esgenoôpetitj who supported their position.

After their experiences of the previous year, non-native residents sought little direct engagement with the native protests. Yet, they were surrounded by the conflict, and encountered it every day. The Burnt Church wharf, located in the English village, remained under occupation. Blockades slowed and prevented travel on local and regional roads. Fishers entered occupied territory every day when they went down to their boats. One fisher said, "For me, since I was at the wharf more than anybody else, there was probably three years like that, arguing, fighting. But I don't mind a good argument" (Mark).

For people who didn't fish, the occupation of the wharf was still an enormous threat: "So when they decided to take over the wharf ... and thousands of native people were there – where did they come from? Who are they? Why are they here? It was like an invasion, of our wharf, our space" (Brenda). When blockades were up, some had to pass through them to leave every morning and to come home every night. Blockades on the main highway through the reserve put residents on their way to the post office or the grocery store face to face with Warriors, and forced them onto long detours. The presence of the Warriors within their village and on the reserve made people very nervous: "Those Warriors that came and took everything over, they were dressed in camouflage and painted their faces ... They were some of the radicals – but they weren't even from here!" (June).[14]

During the dispute, some in the English community felt helpless, defenceless; many felt that their community had been abandoned by other Canadians and they had been left to fend for themselves: "When all this racket was on us, after the cutting was all over, nobody showed up to support, to help or anything like that. We were stuck, this community all by itself" (Luke).[15] As the dispute wore on, the English residents became supporting players, watching events play out between the Canadian government and the Mi'kmaq.

The relationships between the two Burnt Churches, English and Mi'kmaq, remained (unsurprisingly) poor. The United Church minister from the English village worked with the provincial Aboriginal Affairs Secretariat and local leaders to try to bring people together in conversation. But these talks quickly degenerated, and the few people who had attended stopped going. In the First Nation, people were reluctant: "We figured, with all of the animosity that was going [on] that we didn't feel welcome, but if they wanted to meet with the Chief and council, they're our representative ... Throw *them* to the lions and the wolves!" (Barb). Many who did attend felt that their words and efforts were not valued, and so were not interested in going back.

In the English community, people also felt that the talks were going nowhere. Many said that they were hearing the same stories over and over again, and that they couldn't see any way to progress to meaningful solutions. "Some of them have a real chip on their shoulder which they have to deal with ... One, she was saying all of these things that had happened to her family ... She just has to get that chip off her shoulder. I mean, not that the residential schools weren't bad and things, sure they were. But you've got to move on and get it together" (June). People remained unable to talk with one another in a meaningful way – the historical, political, and personal context that was so important to the residents of Esgenoôpetitj seemed an obstacle to the residents of the village of Burnt Church. Like the government, they wanted to get on to talking about how to solve the fishery problem without having to go back and address historic concerns.

As the 2001 fishing season approached, people in both communities felt nervous about what was in store. In the English village, Brenda remembered what the waiting was like: "When the snow melted, you'd have this feeling in your stomach. Okay, what's going to happen this time? Are they going to come back? Whose life is going to be lost?" In the end, the activities of the year 2001 were largely a consolidation of the existing positions. The government enacted more policies and processes to entrench its position in practice and to convey it to the public, and the people of Burnt Church continued to refuse to sign a formal agreement. Overall, there was much less violence reported from Burnt Church in 2001, and with the waning of the violence, media interest also waned.

The DFO and its government partners positioned themselves carefully as the 2001 fishery approached. In February, Dhaliwal and Robert Nault, the minister of Indian affairs and northern development, jointly announced a two-pronged approach to *Marshall* across the Atlantic region (Mi'kmaw territory). First, in an effort to calm the crisis in the fishery, Dhaliwal and the DFO would continue to negotiate agreements with individual First Nations. In April, the DFO and regional native leaders announced that they had been able to agree upon the language for a "template agreement," which would then be used as the foundation for individual agreements across the region. Second, to address the underlying treaty issues, Nault and the Department of Indian Affairs began a process of treaty renegotiation in the Maritimes. Thomas Molloy was appointed as the chief federal negotiator, and these negotiations became known as the "Molloy Process." In early March, the DFO delayed the decommissioning of three Coast Guard vessels in the Atlantic region so that they would be available for enforcement and action in Miramichi Bay should they be needed. Efforts at negotiation were balanced with preparations for direct engagement with the Mi'kmaq.

Entry into the 2001 fall fishery was slightly delayed while the Burnt Church First Nation held its band council elections. Leo Bartibogue, the head of the Esgenoôpetitj Rangers and a prominent activist, challenged Wilbur Dedam for the position of chief. During the lead-up to the elections, there were some allegations that the federal government was exerting undue political influence in the community by calling in third-party auditors to the band council. In the end, dispute leaders who stood for election were largely defeated. Dedam remained chief and regained the majority over council that he had lost in 1999.

One day after native fishers entered the fishery, the DFO announced that it was granting them an eight-day licence to fish for traditional and ceremonial purposes, largely to catch lobster for the annual powwow, they implied. Native fishers and activists argued publicly that the presence or absence of the federal licence made no difference in their ability to fish. The difference it made was to the DFO – if the native fishery was happening under a federal licence, then the DFO could argue that it was unnecessary to take the kind of aggressive and violent enforcement action that they had carried out a year earlier. Non-native fishers in the region did not necessarily concur with the DFO's position, and a flotilla of Acadian commercial fishing boats gathered in the waters off of Burnt Church after sunset one night. In response, native activists again blockaded roads into and through their community.

While the conflict simmered, the DFO renewed its attempts to get the members of the Burnt Church First Nation to sign a fishery agreement. Again, the DFO was attempting to bring the native fishers into its regulatory framework, and again the Burnt Church community insisted that they had the right to fish under their own regulations and conservation plan according to the treaties. Faced with a continuing conflict and the possibility of escalating violence as happened in the previous two seasons, the DFO chose to continue the strategy that it had begun eight days before. It granted a six-week extension to the eight-day licence it had given native fishers. The native fishery continued to be legal in the eyes of the DFO, even if native fishers had not applied for and did not recognize the necessity of this licence.

While tensions remained high throughout the rest of the 2001 fall fishing season, the level of violence in the communities did not approach that of the year 2000. On 11 September 2001, in the midst of the fall fishery, the World Trade Center in New York City was attacked. As the fishing season of 2001 drew to a close, the attention of Canadians was no longer focused on Burnt Church but on New York City, the United States, and Al-Qaeda. The events of 11 September 2001 changed the federal government's approach to policing and protest significantly. Some native leaders and protestors say that at this time they began to be called "terrorists" by some government and media sources; this coincided with increasing government powers of seizure and arrest and intense public sensitivity to security issues. Mi'kmaw leaders, who saw themselves as loyal community activists, found themselves under an exponentially increasing threat. How this specifically affected the events of the rest of that fall, and of the following year, is unclear. It seems likely that they added to the burden of the people of Esgenoôpetitj, began to alter the tenor of debate by casting native protesters as extremists, and added to the threat natives perceived from the Canadian government as the powers available to the government broadened and increased.

The Dispute Subsides?

In 2002, the spring commercial fishery continued, apparently without incident. The media's attention had been drawn away to the "War on Terror" and Burnt Church was left to simmer on a back burner. On 1 August 2002, the Government of Canada's new fisheries minister, Robert Thibault, and BCFN chief Wilbur Dedam announced that they had reached an "agreement-in-principle" that would govern the fishery in Burnt Church (DFO, 2002). Native leaders agreed that the fishers in their community would participate in the spring commercial fishery, and a fall fishery for food and ceremonial purposes, both under the regulation of Fisheries and Oceans Canada. In return, the government provided the community with commercial licences and quotas, boats for the inshore fishery, training for fishers, and funding for fisheries officers from within the community, as well as money to fund studies of the lobster populations. The primary concerns addressed by this agreement were access to, and governance of, the regulated commercial fishery and the traditional and ceremonial fishery; responsibility for the management of lobster populations was clearly agreed to be in the hands of the Canadian government.

Within days, the CPT announced that they would not be returning to the community that year, as the agreement had been signed. Other groups, such as the ARC observers and Warriors from other First Nations, also did not return to the community for the fall season. The conflict had subsided, a temporary agreement had been signed, and the attention of the national and international media focused on other issues.

The *Telegraph Journal* returned to Burnt Church late that summer for one final report. Their reporter characterized the community of Esgenoôpetitj as one divided and exhausted, where many were glad that there was an end to the confrontation but few trusted that the resources that came into the community from the agreement would reach the people in a meaningful way (Klager 2002). The reporter repeated the "hearsay" that the chief and council, corrupt and driven by greed rather than by the will of the people, had lost the confidence of the community.

Many in the community who took great risks during the dispute still felt, in the years afterwards, that their hopes and integrity had been compromised with the signing of the agreement. Many people said to me, "If we had only held on for a little while longer, then the government would not have been able to stand in the face of what they did to us." At the same time, by 2002 some in the community were exhausted, and feeling extremely vulnerable in the post-9/11 political context, where political dissent was seen as highly dangerous. The complexity of the situation is echoed in this woman's reflections on the agreement:

> I have mixed emotions actually, because I think that Wilbur felt he had to do what he had to do. You have to give him credit for hanging in there and supporting the community and going to war with the government of Canada when no other Mi'kmaw chiefs really did that. I think that the pressure just became so great, and I think that he didn't want to see anyone hurt. And perhaps he saw an opportunity to ... make things better. There were those that were tired of the fighting, and were urging him to sign. It was an agreement signed under duress. I know that there were a

lot of us that were not giving up – we just didn't have the power to make somebody not grab a pen and put it to paper. How can you do that? (Alana)

The agreement-in-principle addresses issues of fisheries management, in terms of licences, equipment, and dollars. It does not address the other concerns which motivated this community, deep issues like rights and sovereignty.

Today some people would ask us, if our community feels they won? "Did you get what you were fighting for?" For me, I would say yes and no. Yes because we exposed canada to the world of its racism and injustice against the native people. We also had our story/voices documented so that our next generation will know the native people united in Esgenoôpetitj to protect their rights, as our ancestors did for us. I believe we succeeded in many ways, and the beneficiaries are the next generations. Of course there are also many folks in Esgenoôpetitj that will say that they did not gain nor benefit from the fishery agreement. Many people are still struggling economically and poverty is still an issue. I pray that we will continue to grow stronger and live healthy and well so our children will have a better future. (Miigam'agan)

In a community of great and terrible need, the dispute was a time of great and terrible hope. Few of these hopes were realized in any substantive way. Yet, as time passes, people find that they have not been entirely dashed either. These great hopes are now hopes for the future. Overall, conditions in the community remain difficult. Statistics Canada reports that in 2006 Canada's "poorest postal code" was that of the Burnt Church First Nation. The median taxable income reported by community members was $9200; while band controller Alex Dedam estimates unemployment to be around 45 per cent, councillor Chris Bartibogue argues it remains closer to 80 per cent (CBC 2010).

In the English village, the signing of the agreement meant that the tensions have eased, at least somewhat. People were relieved that the occupation of the wharf, the barricades, and the constant RCMP presence in their community were going to come to an end. But many see that "the root problems are still there" (Paul). People believe that native fishers are still treated differently than non-native fishers, and resent that government hardship monies went to native and Acadian villages after the dispute, but not to Burnt Church. Some see that the increased

commercial fishing in the native community has made a positive difference. And though people insist that the violence of the dispute will not
happen again, in their hearts, some remain worried. "I'm sure that each
person in this community still has dreams that scare them. It could
happen again, and if it does happen again, it's going to be worse"
(Brenda).

"The Root Problems Are Still There"

Paul is a careful, thoughtful man who is also cautious. His analysis that
"the root problems are still there" comes out of his own long consideration of the events in his community. He goes on to observe:

> But really, the fishery issue created a lot of problems, and – most people
> will die in this community without any great deal of empathy for the na-
> tives. They're just too old and they don't, they're not gonna change. I
> guess all you can do is hope that the younger generation coming up is
> going to be a little more open minded … [but] there's not so many [young
> people]. In this community, anyway.

In Paul's view from within the English village of Burnt Church, the
estrangement of these two racialized communities is significant. The
inability of people to hear one another, to see one another, and to have
the courage to address one another across the lines which divide the
English and the Mi'kmaw is a stubbornly persistent part of daily life.

The estrangement between the English and Mi'kmaw communities
has roots far deeper than the dispute. The two communities hold differing views of the history of the dispute, and its origins, and have had
very different experiences of their government throughout their his-
torical·lives. But we can also begin to see, here, that these two communities have different senses of relationship to the past, and different
experiences of history. While those in the English village see the events
of the last six or seven decades as relevant to the evolution of the native
fishery, many also believe that the past is something that should be over
and done with. The conflicts and promises of the generations that came
before them are important for their historic value, but are not alive to
them in the present day. In contrast, many Mi'kmaq understand the
historic treaties as living documents which should be adhered to in
present-day relationships, and see the movement away from those
commitments not as a historical failure but an ongoing betrayal of trust.

For the Mi'kmaq, historic agreements and the stories and insights of their elders and ancestors remain central to the lived experience of the present.

Heidegger suggests that our existence is necessarily historical, and that, as such, "the true object of historical investigation is not the facts of a past era, but a possible mode of existence" (Mulhall 2005). In *Being and Time*, he writes that each human being

> *is* its past, whether explicitly or not. And this is so not only in that its past is, as it were, pushing itself along "behind" it, and that Dasein possesses what is past as a property which is still present-at-hand and which some-times has after-effects upon it: Dasein "is" its past in the way of *its* own Being, which, to put it roughly, "historizes" out of its future on each occa-sion ... Its own past – and this always means the past of its "generation" – is not something which *follows along after* Dasien, but something which already goes ahead of it. (1962, 41; emphasis in original)

The past is not a neutral, atemporal set of facts. History is not a creature of reason. History is a practice of constructing self and community, the way in which the present negotiates itself into the future. In this sense, history is foundational to the nature of human Being. Confronting the colonial past in the settled present doesn't just challenge the facts of history as a field of study; it challenges the settlers' visions of them-selves as peaceable Canadians, their sense of their own legitimacy and identity. The anguished protests of the English villagers are existential cries, attempts to defend the possibility of a future in the face of the past:

> I think it's a terrible thing, what was done to the Indians at the time – but I had nothing to do with it! And they had nothing to do with it! They weren't even involved when the British government decided to take them, put them all on reservations! That should never have happened! But it did happen. But how can I be held responsible for what they did? ...
>
> And how can they feel so terrible for what happened to their ancestors, way back when? There's nothing they can do about it, any more than there's something I can do about it! I don't understand it! (Mary)

The trauma of historical re-membering, of the recognition of the mutal-ity of colonial relationships, is a great burden. The oblivion of the colo-nial aftermath protects us not simply from the facts of our shared past, but from the troubled nature of our existences. Addressing the past can be terrifying.

The task of addressing the past is existential; it is also political. Foucault recognizes "the power and usefulness of seeing our present world in terms of the way that the past has shaped us" (Mugerauer 1994, 29) and emphasizes, in this context, the determinative power of over-arching social structures. In his later work Foucault argues that the structures of power of the state are revealed through an analysis of governmentality,

> the ensemble formed by institutions, procedures, analyses and reflections, calculations, and tactics that allow the exercise of this very specific, albeit very complex, power that has the population as its target, political economy as its major form of knowledge, and apparatuses of security as its essential technical instrument. (2007, 108)

In Burnt Church, the profound differences in the Mi'kmaw and English experiences of government and of governmentality predate the earliest contacts and settlements, and extend through colonial settlement into contemporary federal projects of management, regulation, and enforcement. These historical experiences of governmentality shape the existence of the two communities in the present day, as I will explore in the coming chapters.

The eruption of the dispute betrayed the tenuousness and tension of many relationships. In Burnt Church, it revealed a depth of conflict and pain in relationships between people that had long been unacknowledged. The dissociation of the two Burnt Churches from one another extends through differing experiences of history, and differing experiences of government, into the senses of agency and community at play within each group. The Mi'kmaw and English residents of Burnt Church also talk about the importance of understanding their relationships in and with this *place* to understanding the dispute. Chapter 3 explores place-based identities in historic and contemporary Burnt Church, and the development of Burnt Church/Esgenoôpetitj as a contested place.

3 Contested Place

If place-making is a way of constructing the past, a venerable means of *doing* human history, it is also a way of constructing social traditions and, in the process, personal and social identities. We are, in a sense, the place-worlds we imagine.

(Basso 1996, 7)

The relationships and disjunctures between the Mi'kmaw and English Burnt churches are written in landscape. The two communities are spread out along the water where Church River runs into Miramichi Bay. They are the northern and easternmost communities in the region known as the Miramichi, located on the boundary where the Miramichi meets the Acadian Peninsula. While the province maintains the roads in the English community, responsibility for roads has been downloaded from the federal government to the band council in Esgenoôpetitj, who have struggled with the task. So in my first days in Burnt Church, it became clear to me when I had moved from one community to the other, simply because of the grey line in the road where the new asphalt ended.

After a couple of days in Burnt Church, the differences among the homes on the two sides of that line also become apparent. The English village is dotted with older farmhouses and newer bungalows, the homes of the year-round residents nestled among small fields of agricultural land and forests. The grand old houses along the edge of the bay are now used as summer homes by those who have moved away. In Esgenoôpetitj most people live in government-built bungalows, some new, some terribly old and run down. After a few months in the communities, someone pointed out to me that the boundaries between the communities are also marked by red roadside posts, presumably erected by the Department of Indian Affairs in a previous era. The posts are now hidden in the overgrowth at the sides of ditches, but they remain as warnings to those who know to look for them, physical markers of the break between the two communities.

During my year in Burnt Church, I had the opportunity to participate in hosting other outsiders who, like me, came to the First Nation interested in what happened in the dispute, and in its aftermath.[1] At these times, a small group of people from the community welcomed the visitors and shared some stories from the dispute with them, as they had done with me when I first arrived. As on my first trip to the community, they shared their stories by inviting the visitors into vehicles and escorting them around the community to see how and where the dispute had occurred. The tours stopped on the boundary of the community where the red posts stand and paused under the water tower, where the Mi'kmaw name of this community, Esgenoôpetitj, is written. People would get out of their cars at the bridge over the river where the dories are pulled up, and walk around at the site of the sacred fire on Diggle Point where the powwow takes place. The tours finished along the shore so that visitors could see where the dories had fished during the dispute; where fishers had encountered violence with the government and their neighbours; where people stood on the shore watching and praying for their family members; where the news trucks pulled up with their satellite dishes and cellular phones. The wharf, which was occupied by members of the First Nation during the dispute, is pointed out – past the posts – in the English Burnt Church.[2] And then the group would typically return indoors, to someone's home, where a discussion could develop in more comfortable surroundings.

Sharing the story of the dispute was not separate from sharing the place in which it happened, even in the middle of winter. Telling the stories of the dispute, even briefly, seemed to require being on the land and at the water's edge, in the places where these events had played out. As outsiders coming into Esgenoôpetitj, these people (including myself) were asking about events, about ideas and actions. Yet, in response to these overtures, place came first. What is being said, implicitly, is that the best way to understand something about the dispute and the people who lived it is to come and see the place where it happened. Once you have seen what is there, then maybe something else can be shared; but until one has experienced the place, little of the rest of the story will make sense.

In the English community of Burnt Church, among the descendents of settlers, conversations about the dispute also led to conversations about place. I often asked people if there were things that they had stopped doing because of the dispute. In the English community, nearly everyone felt less safe to move freely about the village. Many people, especially women, take walks in the morning or evening as exercise and social time. During the dispute, people feared for their safety so greatly that many (but not all) stopped travelling the roads and lanes on foot and stopped driving through the First Nation.

> It wasn't safe here. We couldn't even go walking. My daughter had to stop walking down the side of the road in our own community. The wharf was our wharf, part of our community and what we built and took care of, and we couldn't go on it. It wasn't safe. (June)
> A lot of people, to this day, are afraid to drive through the reserve ... when the fishing was going on, then I didn't, because I was afraid. (Mark)

By the time I arrived in the community, and the dispute had subsided, people had begun to walk again. Some told me that they had driven through the reserve for the first time since the dispute some two years later. And others lamented the change in the social fabric of the community. Before the dispute, the wharf had been a place for people to congregate on warm evenings, as a part of their constitutional. For generations, the wharf was at the social, cultural, and economic core of the community. After the Mi'kmaw occupation of the wharf during the dispute, that changed. Today, it remains a place of work and busyness, but is rarely the destination for a stroll as it once was.

The English people of Burnt Church have become tied to these lands and waters on the edge of Miramichi Bay. Most can still trace their ancestors back to those who arrived in the community with land grants given them by King George III. They are at home in Burnt Church in a way that they can experience no other place; this is their place in the world. Yet in their own homes, they experienced the dispute as a violent and destabilizing event, one in which their neighbours asked them to be accountable for the actions of these same ancestors. Brenda lamented to me, "Why do I have to pay for the sins of my forefathers?" The dispute raises the spectre of colonialism, as the legitimacy of settlement and livelihood is questioned both directly and indirectly by activists engaged in it.

The anthropologist Keith Basso explores the significance of places to the life of the White Mountain Apache in present-day Arizona in his remarkable work *Wisdom Sits in Places* (1996). Basso was invited by his friends and colleagues among the Western Apache to work with them in developing maps and accounts of their traditional places and place names. He argues that "Apache constructions of place reach deeply into other cultural spheres, including conceptions of wisdom, notions of morality, politeness and tact in forms of spoken discourse, and certain conventional ways of imagining and interpreting the Apache tribal past" (xv). Detailed conversations about place, over many years, led Basso into new understandings about many dimensions of Apache life. As Basso points out above, this is because place itself shapes and is shaped by the cultural lives of people.

In Burnt Church, place is not explicitly important to outside conceptions of the dispute. The dispute seems to be about lobsters, quotas, tags, and boats; at a deeper level, we might recognize that the dispute has something to do with race, power, and violence, with colonization and conceptions of justice. But my conversations and relationships with people in both Burnt Churches, in which we talked about these issues in depth and at length, often seemed to lead to place. Unlike Basso, who set out to learn about place and found himself learning about many other things, I set out to learn about other things and found myself learning about place. When I commented on this challenge to Leo, who lives in Esgenoôpetitj, he laughed and said, "Sometime you'll figure out what matters here, eventually. If you pay attention." Discussion of place is necessary to understanding the perspectives of local Mi'kmaw and English residents and their characterizations of the events that began after the *Marshall* decision in 1999. In Esgenoôpetitj, place is both the context of the dispute and its focus.

When I first visited the Burnt Church First Nation, or Esgenoôpetitj, one of the people to spend time with me was Lloyd Augustine. Kwegsi, as he is known in Mi'kmaq, is the traditional chief in the community (keptin in the Mi'kmaq Grand Council, or Sante Mawi'omi)[3] and a carpenter. As outlined in chapters 2 and 5, during the dispute Lloyd worked with James Ward to carry out a community consultation process and write the community's Fishery Act and Management Plan (Ward and Augustine 2000a, 2000b). We spent many hours in his living room, surrounded by cats and dogs and children, as Lloyd tried to educate me. What was important about the dispute, for Lloyd, was that it was an expression of the historic concerns of the Mi'kmaw community as they have developed throughout the colonization of their territories. Lloyd argues that the land on which Canada is built is not rightfully Canadian but Mi'kmaw and has never been ceded:

> Whatever I keep from taxes and resource revenues from the Canadian government, it's actually a pittance of what rightfully belongs to me. There's no paperwork that's there where we have handed over to them the deed to what is there … We have always believed that what is there belongs to the Creator and cannot be sold or given up.
>
> White people's anger stems from the idea that they are dealing in stolen goods. They tried to terminate or exterminate or assimilate the Indian … but their own guilt makes them realize that no matter what they do … it'll always be to their own shame.
>
> Even if they wiped us all out, the children looking at history books will always question, "Who are these people? Why did they die?"
>
> "Because they thought this was their land."
>
> "Why did they think that?"
>
> "Because they were here first."

For Lloyd, this historic sovereignty is at the heart of the conflict between First Nations people and Canadian settlers, between his people and their neighbours. The lands and waters, in his sense, are not something that can be owned and parcelled out. The idea that land's value comes from the labour that people put into it, as John Locke suggested and the British Empire understood, is in fundamental opposition to Lloyd's position. Traditional Mi'kmaq philosophy understands "what is there" to be a sacred trust given to the people by the Creator, something that cannot be given up. The carving up of Mi'kma'ki into properties for settlers results in a conflict that's not simply about ownership, but about the fundamental nature of the place and human relationships to it. While some of these philosophical differences will be discussed in more detail in chapters 4 and 5, taking them up here illustrates that place is important as the subject of the dispute, as well as being something that shapes it.

As Basso argues, place and identity are "inseparably intertwined" (Basso 1996, 35). The geographic and cultural relationships that arise in a particular place shape who we are as individuals and as communities of people. For Basso, whose thinking is influenced by Deloria, this is true in much of indigenous North America. "The past lies embedded in

. features of the earth … Knowledge of places is therefore closely linked to knowledge of the self, to grasping one's position in the larger scheme of things, including one's own community, and to securing a confident sense of who one is as a person" (34).

This understanding has echoes in the phenomenological literature on place, where place is understood to be fundamental to the identity of all persons, even in cultures where this reality remains hidden or obscured. The philosopher J.E. Malpas argues that "the landscape in which we find ourselves, and through which we are defined, is [thus] as much a part of what we are, of our minds, our actions, and our selves, as is the food we eat and the air we breathe" (1999, 189). The invitation into the places of Burnt Church is, then, an invitation into a conversation about the identities of those who live there, into their minds, actions, and selves.

The construction of place and of identities in place (which is perhaps the same thing), while clearly geographic, is also historical. Stefanovic reminds us that "history is not something that we own, but something that we *are*" (2000, 107; emphasis in original). The ways in which we do, and know, and understand our own histories shape us; so do the histories of the places which are our homes. We build places with the past, creating their meaning and significance through that which came before us. Basso suggests that places are not just the "fleshing out of historical material … Place-making is also a way of constructing history itself" (1996, 6).[4] Place, history, and identity are mutually constituted at the level of being. They are fundamental aspects of existence, the interconnected elements not simply of particular philosophical systems, but of self and community. In this context, my attempt to articulate Burnt Church/Esgenoôpetitj in these pages, and to convey how this place shapes identity and experience in the present, must attend to history as an existential element of the dispute.

Getting to place takes us to the heart of the dispute in the Burnt Churches. Burnt Church (English) and Esgenoôpetitj/Burnt Church First Nation (Mi'kmaq) are paradoxical places. They are communities deeply separate and distinct, even in opposition to one another, yet shaped by a shared colonial history that is centuries old. In both communities, the very sense of place that shapes community life and identity is grounded in the experience or threat of displacement. Examining the historical, geographic, political, and spiritual construction of this place demonstrates that danger and uncertainty are integral to it. Burnt Church is a contested place, two communities in which the tensions between people's profound and long-standing relationship to place and the profound terror of loss of place forge community, identity, and landscape.

A Short History of the Burnt Churches

In the Burnt Churches, the Mi'kmaw and English communities live side by side, but their social and cultural lives are almost entirely separate. Neighbours from the two communities know one another, and one another's family histories, but rarely socialize or work together. They are surrounded by Acadian villages, the largest of which is Néguac, which lies to the north. The distinctions between these three communities are strong and fast and have held for generations. Simply by knowing a person's last name in these communities, you would also know their language and culture (English, French, Mi'kmaq); their religion (Protestant, Catholic, traditional Aboriginal); their political affiliation (Liberal or Conservative), since politics are usually familial; and perhaps even the lands on which their family lives.

When I arrived in Burnt Church, people were interested to know my last name and who my father was, so that they could place these things about me. Since I share a last name with a local English family, there was often some interest in whether I was a long-lost relative. Even though that was not the case, my English neighbours were not wrong to deduce that, with the last name King, I come from a lineage both English and Protestant. The fact that I came *to* Burnt Church, rather than moving *away* (as most young people have had to do, for work) also made me something of a rarity in the English community. The First Nation has much more of a history of non-native missionaries, academics, and "do-gooders" showing up for periods of time, and I gather that I was seen to fall into this group, for the most part.

The contemporary dynamics of relationship and estrangement among the English, French, and Mi'kmaq at Burnt Church have their roots in the past. The Mi'kmaw people of Esgenoôpetitj thrived for centuries, fishing and hunting, in communities of great numbers before the epidemics associated with European contact of the sixteenth and seventeenth centuries. This pre-contact history is a rich and important time for Mi'kmaw people, something that Mi'kmaw community members and scholars are working to understand and to reclaim in the present day. But it is the history of contact and colonization – with the ongoing conflict between French, English, and Mi'kmaq – that is critical to this account of community relationships. The first contact that Mi'kmaw people of the Miramichi region had with European colonists was likely with European fishers in the mid-sixteenth century. In the 1620s, French traders built a post on the island of Miscou, an easy trading distance from Miramichi Bay, and about a decade later Jesuit missionaries began to make visits to the area, some of which are documented in the *Jesuit Relations* (Basque 1991, 27). The Récollets returned to New France in 1670 and took on Miramichi as a mission field by the end of the seventeenth century (44). Christian missionaries built relationships and a stone church at Esgenoôpetitj as the eighteenth century began (53). This church gave Esgenoôpetitj one of its first colonial names – Pointe-à-l'Église, or Church Point.[5]

Esgenoôpetitj was embedded within Mi'kmaw territory. Mi'kmaw people hunted in family groups during the winter, and then gathered together in extended social and family groupings at summer gathering places each year (Wicken 2002). Today, the Mi'kmaq of Burnt Church recognize that Esgenoôpetitj was one such summer gathering place. Wicken suggests that Mi'kmaw people understood themselves as belonging to a larger social and cultural group, what we now call the Mi'kmaw Nation, whose territory stretched from Gaspésie into present-day New Brunswick, Nova Scotia, Maine, and Newfoundland. In northeastern New Brunswick, the pre-contact economy of the Mi'kmaq was based largely on the spring and summer fish runs; winter hunting happened more or less in the same geographic areas (Nash and Miller 1987, 48). Because of the abundance of fish, the Mi'kmaq who lived along the Miramichi, and especially along its lower portions (where Esgenoôpetitj is), developed an economy highly adapted to fishing (48).

While Mi'kmaw life and custom continued during the early years of contact with the French, cultural change was already afoot. The Mi'kmaq who lived in closer contact with the Acadians were economically and spiritually influenced by their new neighbours. On 24 June 1610, Chief Membertou and his family were baptized at Port Royal, the first Mi'kmaq to be baptized in Mi'kma'ki (Prins 1996, 81).[6] Baptism functioned as a way to cement alliances and relationships between the Mi'kmaq and the French in those early days (Prins 1996, 80–2; Upton 1979, 20–5). Missionaries also began to be seen as skilled in the arts of prediction and medicine, and their usefulness to the Mi'kmaw grew accordingly (Upton 1979, 22). Over the next century and a half, the Mi'kmaq continued to develop their own understanding and practice of Christianity, and by the end of that time many Mi'kmaq clearly also identified as Catholic (65–9). Catholicism became a Mi'kmaw religion, one which the Mi'kmaq continued to practise after the ascendance of the British and the departure of French Catholic missionaries from Mi'kma'ki (Upton 1979, chap. 11, 154).

In 1722, the Mi'kmaq and their allies (taken together, the Wabanaki) entered into a war with the British colonies at Massachusetts over land (Wicken 2002). This was the only one of the Anglo-Wabanaki wars that was "a local war, rooted in Indian reaction to British intrusions" (Prins 1996, 137). By 1726, all the Wabanaki allies excepting the Mi'kmaq had signed treaties with the British. Later that year, the Mi'kmaq did the same, signing their first Treaty of Peace and Friendship with the British Crown. Much of this action, while involving the Mi'kmaq as a whole, took place in the southern parts of Mi'kmaw territory. The contestations between the British, the French, and the Wabanaki were focused in New England and Nova Scotia/Acadia and only moved into the Miramichi later.

In 1755, the ongoing violence between the French and British over their colonies in the New World took an important turn. The Acadians were expelled by the British in what is still known as Le Grand Dérangement. Many Acadians were sent to other colonies; many also stayed on the Maritime coast. Of these, some moved to existing settlements and embarked on transient lives, waiting for the time to return safely to their homes. Others moved north, into the coastal woods of present-day New Brunswick, seeking shelter from which to resist the British.

By the winter of 1756–7, some Acadian resisters had arrived along the Miramichi, taking shelter in the woods, building alliances, and planning raids on the British with the Mi'kmaw people who lived there.[7] The first winter that the Acadians spent in northern New Brunswick was brutal. They lived on the very edge of survival, and those who made it through only did so with the help and aid of the Mi'kmaq, who showed them how to survive in the frozen forest. Yet joint Mi'kmaq-Acadian attacks on British ships on the Gulf of St Lawrence continued.

In the fall of 1758, frustrated by ongoing losses, British General Wolfe ordered James Murray to remove the final pockets of Acadian resisters who had taken refuge on the Miramichi. Murray targeted the settlement at Esgenoôpetitj/Pointe-à-l'Église:

> I therefore in the evening of the 17th in Obedience to your instructions embarked the Troops, having two Days hunted all around Us for the

Indians and Acadians to no purpose, we however destroyed their Provisions, Wigwams, and Houses, the Church which was a very handsome one built with stone, did not escape ... and I am persuaded that there is not now a French Man in the River Miramichi, and it will be our fault if they are ever allowed to settle there again. (Ganong [1914], in Basque 1991, 55)

In fact, the Acadians, the Mi'kmaq, and the missionary had all taken to the woods, where they hid successfully until the departure of the British. They returned to find their homes destroyed. The Acadian settlers eventually moved north and built the new settlement of Néguac. In 1760 and 1761, Mi'kmaw leaders across Mi'kma'ki and in Miramichi renewed treaties of peace and friendship with the British, treaties that became the basis of the *Marshall* decision in 1999.

After a period of about thirty years, settlers arrived near Esgenoôpetitj under the auspices of King George III. A colony developed west of the Mi'kmaw community at the old Church Point site, now known as Burnt Church. The settlers, who held King's Grants to their lands, were mostly Gaelic-speaking Scots Presbyterians, many of whom had settled earlier in other communities on the Miramichi River (Wasson and Murdoch 1999). Most people who live in the village of Burnt Church today are descendents of those original grantees, or of two or three other English-speaking families who arrived at the same time. Some families still have the original grant papers that awarded them their land.

Local historians suggest that their ancestors saw echoes of Scotland in their new home, a new place which welcomed them because it echoed the old.

These settlers took advantage of the resources available from the sea, the forests and small farms. The location was perhaps not too different than that with which they were familiar in Scotland: the incoming tides, the smell of moist air, heavy with salt and the northeast wind. They brought with them their Scottish heritage, a sense of duty, a sense of community and their religion. (Wasson and Murdoch 1999, 2)

Ganong ([1914], cited in Wasson and Murdoch 1999) notes that although the settlers from Burnt Church were closely tied to settlers in other British communities, Burnt Church was distinguished by the absence of Loyalists and Revolutionary soldiers among the settlers. As the community grew, the next generation of descendents also took up

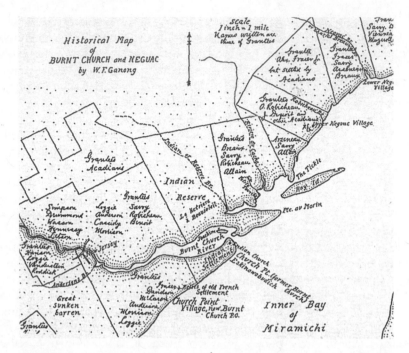

Figure 4. Nineteenth-century Burnt Church and Néguac, indicating Burnt Church Indian Reserve, Burnt Church settlement, and surrounding settlements. The names on the map are those of the families who were individual grantees in the original settlements. Note that the reserve lands include the "settlement" at the end of the peninsula, as well as lands directly to the north. The Tabisintac River is not shown. (Map drawn by W.F. Ganong [1908]; in Basque 1991)

grants, further upriver at a settlement named New Jersey. The ties between these two communities, and the strong family connections, remain to this day.

The success of the settlements at Burnt Church began to put pressure on the Mi'kmaq. In 1801, they petitioned for a "License of Occupation" granting them permanent access to their eeling grounds and preventing further settlement on specific territories. In 1802, a reserve was granted of 9035 acres on Burnt Church Point, with 1400 acres adjoining on the north side of Church River, and a second location of 9035 acres at the eeling grounds on the Tabusintac River (Upton 1979, 99). Given the enormous impact of disease on indigenous populations, including the Mi'kmaq, it is difficult to determine what the pre-contact population of the community was. Miller suggests that the pre-contact population across Mi'kma'ki was around 35,000 (1976). The entire Mi'kmaq community was decimated by disease, across Mi'kmaq territories which came to be called New Brunswick, Nova Scotia, and Prince Edward Island by colonists.[8] In 1841 in the province of New Brunswick M.H. Perley, commissioner for Indian affairs, reported that the total Mi'kmaw population was around "935 souls" (Perley 1841, 82). Perley further reported that the impact of disease was tremendous, and "while [I was] at Burnt Church Point, a death occurred almost daily" (83). By the turn of the twentieth century, the community at Esgenoôpetitj was about 200 people (Martin, n.d., unpublished paper).

By the time of my arrival in Burnt Church in 2004, the reserve population had grown to about 1300 people, many of them children. Housing and employment are huge issues for the members of this community, where the only local infrastructure is that supported by the band council. The community school stands on the shore, across from the Adult Education Centre, which also houses the band council. Beside the school is a health centre, and there is a day-care and early years centre as well. St Anne's Roman Catholic Church stands beside this small complex of band council institutions, a small white wooden church maintained by community members; attended by a loyal group every week, it is full on holidays. The elected chief runs a small grocery and gas bar out along the highway. There is no public wharf in Esgenoô-petitj; Aboriginal fishers who have fishing boats fish from the wharves in Burnt Church, Néguac, and further north at Tabusintac. Many community members still keep small dories, pulled up inside the mouth of Church River where it meets Miramichi Bay.

The other important community space in Esgenoôpetitj is Diggle Point, the location for the community powwow every August. An arbour is erected there, beside the place where the fire-keeper guards the sacred fire throughout the powwow. During most of the year, Diggle Point is something like, but more than, a community park. And for a weekend in August, community members, friends, and relatives gather there, with tents and campers, for festival and celebration. Esgenoô-petitj/Burnt Church First Nation has many community members working for the revival and healing of the community. Some of these people came together to work publicly for change during the dispute. Now others work quietly within their families to provide hope and support, keeping space for change.

By the early twentieth century, the settler village of Burnt Church was prospering. The shore was lined with businesses – a store, a lobster canning factory, and a small fleet of boats for inshore fishing. The Presbyterian church built by the first settlers was replaced with a larger church building with a tower. In 1925 the congregation voted to join the union of the United Church of Canada. A one-room school was built not far from the church, and the Women's Institute raised a hall to host their meetings and many other community functions. The lobster fishery was *not* highly lucrative throughout most of those years. Lobsters were used as fertilizer and then as food when times were tight, only becoming profitable as lobster became a luxury good in the latter third of the twentieth century. People supported their families by fishing a variety of stocks, including eels, oysters, herring, mackerel, and crab. Even with the upswing of the lobster fishery, the economy of the late twentieth century did not favour small fishing communities, as can be seen across the region. Independent inshore fishers in small boats were being out-competed by larger offshore enterprises, squeezed out by increasing corporate and multinational dominance in all fisheries, and finding their livelihoods jeopardized by the decreasing catches as stocks were "fished out." Eventually the factory and store closed. The tiny local credit union is the one remaining storefront in the community.

When I arrived in the summer of 2004, the English community of Burnt Church numbered about eighty-five persons, though the population swells for a short time every year with the arrival of the "summer people" who own cottages or visit the tourist hotel along the shore. Most year-round residents live along the Upper Road, where St David's United Church and the old Burnt Church School (now the Seniors' Centre) are also located. The Women's Institute Hall serves as a community centre for meetings, dances, and meals. In the late spring, the wharf is a bustling place, as local fishers make the most of the lobster season. The community remains committed to its own survival. In the year that I was there, committees of local people built a cenotaph,[9] fundraised to host a Canada Day festival, raised money to maintain the community buildings, and continued to manage the public wharf and the nine-hole golf course. Life in this community is rich, grounded, and tied to this place and the stories of the early settlers. And yet, the ongoing conflict with the people of Esgenoôpetitj makes the double nature of this place once again visible. This small fishing village is like many other

Canadian places, robust in its own terms yet fundamentally destabilized by conflict with its indigenous neighbours.

The Acadian residents of Néguac have built a thriving village of 1600 people in the long aftermath of their expulsion. For many years, Acadian people were treated as second-class citizens by a New Brunswick government (and populace) dominated by English power structures. (Aboriginal people were seen as a distant third priority, behind these two other groups.) After the election of the Robichaud government to the provincial legislature in 1960 – and the implementation of the "Equal Opportunity Program," designed to eliminate the disparities between English and Acadian communities across the province – the position of Acadian communities and people in the province rose dramatically. Néguac is the local commercial centre for the Burnt Churches as well as for the Acadian community, home to the local drugstore, hardware store, lumber yard, Tim Hortons, post office, grocery store, and dépanneur (though people also travel to Miramichi regularly to the larger grocery store and the Walmart). For people in both Burnt Churches, running to the corner store or for coffee means driving the five to fifteen minutes to Néguac, and some from Néguac attend church at St Anne's or square dances at the Women's Institute Hall. Lives in the Mi'kmaw, English, and Acadian communities remain as they have always been – separate yet intertwined. The alliances, conflicts, and colonial politics which shape the history of this place are reflected in the social landscape of the present. This place is built with and from its past.

Religion and Place

Considering place requires understanding not only the social landscape of the past, but also the social and cultural dynamics of the present. Religion is sometimes seen as irrelevant to environmental conflicts – yet many such conflicts can be understood as place-based conflicts, and religion is critical to place. Anthropologists of place consistently recognize the importance of religion to their studies. Low and Lawrence-Zuniga's *The Anthropology of Space and Place: Locating Culture* (2003), and Gupta and Ferguson's *Culture, Power, Place* (1997) engage with religion as a fundamental part of culture, identity, and place making. Tim Ingold's work on the anthropology of dwelling and livelihood discusses the significance of understanding "cosmological conceptions of the earth" in this context (2000, 153) and forms part of a fascinating volume, *Imagining Nature: Practices of Cosmology and Identity* (Roepstorff, Bubandt, and Kull 2003) that explores the significance of nature as "simultaneously semioticized and real" (Roepstorff and Bubandt 2003, 26). In Burnt Church/Esgenoôpetitj, religion is an obvious issue from the outset. In a place named *Burnt Church*, how are churches – and religion, writ large – important? Addressing place names yields insight into the importance of religion as a historic factor, mediating and shaping people and place. But religion is not just something of the past; it remains an important contemporary factor. In the face of the dispute, people try to understand the meaning of their lives, communities, and situations, and this impulse, whether lived out in organized religion or in some other form, is a religious impulse. The communal sensing and expression of place is in itself a religious exercise (Basso 1996, 145, 148), the conscious or unconscious negotiation of the grounds of existence. If we are to understand the nature of people's relationship to place, then we must consider this relationship in all its dimensions – historical, political, colonial, geographic, *religious*. The notion of place is an idea, in part, about religion and the religious.

At the 2007 annual meeting of the American Academy of Religion, a session of the religion and ecology group focused on place. The papers and conversations in "Religion from the Ground Up: Religious Reflections on Place" characterized place as an essential notion that should be reclaimed and re-inhabited as a way to resolve the disenchantment of the world.[16] Dr Stephanie Kaza, who responded to this session, asked what more there might be to the notion of place. Is it simply that we all should rediscover a sense of place and the world would be put in order? What about places in conflict, or where different people have different senses of place? Thinking of sense of place as a panacea, as something that needs to be reclaimed in an uncritical exercise, collapses our conversations about the religious dimensions of place into devotional ones. People already exist in places and have senses of place that need to be examined and considered. As Mick Smith argues, reclaiming place does not necessarily contribute to a better environmental ethic or world view (2001). Exploring the religious dimensions of place, as is done here, is not a prescriptive or theological exercise, but one that seeks to understand people, landscape, and culture. Developing a *critical understanding* of place and its religious dimensions may help us to approach some practical solutions to social and environmental problems, as problems of relationship to human and non-human beings.

In the case of Burnt Church, we face a challenge clearly environmental and political. Yet some of the implicit, historical, and cultural dimensions of this conflict, and of this place, are religious. Religion is, after all, embedded in the name of the place! Casey points out that the names of places themselves embody and instantiate a community in its historic and geographic setting, indicating something about how people and landscape are, together, socially implaced (1993, 23). In his work with the Apache, Basso furthers this argument, suggesting that

> place-names are among the most highly charged and richly evocative of all linguistic symbols. Because of their inseparable connection to specific localities, place names may be used to summon forth an enormous range of mental and emotional association – associations of time and space, of history and events, of persons and social activities, of oneself and stages in one's life. (1996, 76)

In Burnt Church/Esgenoôpetitj, the name points us towards the religious, colonial, and conflicted nature of this place. Before colonization, this place was known as Esgenoôpetitj, "look out place where one waits for the others" (Martin, n d , unpublished paper), or "the People watching for those to come" (gkisedtanamoogk 2007). Esgenoôpetitj was a summer camp and gathering place for Mi'kmaw people. Before and at the time of contact, "Mikmaq place names had more than a functional use; they also strengthened a community's identity with the surrounding landscape" (Wicken 2002, 37). Contemporary Mi'kmaw use of this traditional name (as is common practice in Mi'kmaw communities across Mi'kma'ki) ties this community to their ancestors, to times and lives before colonization, and to a Mi'kmaw world view that lives contemporaneously alongside and within Canadian society.

The name Burnt Church evokes the conflict of colonialism, the war with the English, and the importance of religion as an element of the colonial encounter. In New France, it was commonplace for relationships with indigenous communities to be built by missionaries, such as those who came to Miramichi with the support of the local seigneur, Nicolas Denys. With the arrival of these missionaries after the late 1600s, the conversation, confrontation, and alliance between Mi'kmaw tradition and Catholicism became a feature of life in Esgenoôpetitj. The stone mission church at Esgenoôpetitj/Pointe-à-l'Église was burned in the attack on the Mi'kmaq/French alliance by the English in 1758. The

name Burnt Church marks not only the importance of the alliance between the Mi'kmaq and the French, and of the church to that alliance,
but the threat and conflict under which the Mi'kmaq and French lived
and the victory of the British forces. It is an English name for a community originally Mi'kmaq, and then also French. As the British and their
Canadian descendants retained the power of naming the First Nation
(The Burnt Church First Nation) and the settler community (Burnt
Church), perhaps it is no surprise that a name marking an apparent
British victory was chosen. Bound up in an English name about victory
and defeat are the ghosts of alliance, and the spirits of inter-religious
encounter, at the heart of an early colonial community.

Place names remain contested in both Burnt Churches to this day, as
people negotiate the representation of their identities. In the Mi'kmaw
community, though the government calls them the Burnt Church First
Nation, many people also continue to call their community Esgenoôpetitj. Esgenoôpetitj is painted in bright letters on the community water
tower. Like most other Mi'kmaw communities, the people of Esgenoôpetitj are working to retain their language and cultural identity, and
retaining their own name for themselves is an important part of this –
and also a political statement. In the region, though, non-Mi'kmaq often do not recognize the Mi'kmaw name, and so people from the reserve
comfortably use both, using one or the other more commonly depending upon their politics and comfort in their language.

The racialized nature of the name Burnt Church is more complicated
regionally. An academic colleague of mine who grew up in the
Miramichi tells me that at his high school, Burnt Church was thought to
be so named because "the Indians burned the church." A name that re-
calls the burning of a French and Mi'kmaw church by the English in a
violent colonial relationship has become, instead, another cipher for the
"Indian problem" (see Dyck 1991). The name echoes and reinforces
racialized stereotypes about violent Indians and victimized Christians
as separate and opposing groups, a process that was amplified and extended by the violence of the dispute itself (and representations of
this violence in the media). Perhaps these local (mis)interpretations of
the name Burnt Church further illuminate the importance of the name
Esgenoôpetitj for the people of the First Nation.

In the English community of Burnt Church, people considered changing the name of their community as the dispute wore on, perhaps to
Church River (after the river that divides the peninsula from the surrounding communities). As many of their neighbours in the region

believed that Burnt Church was only an Indian community and not also a white settlement, local whites felt invisible, or tarnished by the name. A small community of only eighty-five persons, the English felt maligned and ignored in the media coverage of the dispute. They held a referendum, which did not pass, and the name of the community remained as it has since the days of General Wolfe. In the present, the name Burnt Church seems to have little to do with factors related to religion. In fact, the name itself demonstrates how complicated and interconnected religious life is with the rest of everyday life, with the historical, political, colonial, and racialized construction of Burnt Church/ Esgenoôpetitj.

Religion remained an important historical factor in Burnt Church throughout the eras of colonization, both for the English settlers in Burnt Church and for the increasingly pressured and marginalized Mi'kmaq. People's relationship to place and to one another in place can be expressed through myth and ritual, music, art, and prayer, weaving together social relationships and refastening them to the landscape (Basso 1996, 109–10). In the developing communities of Burnt Church/ Esgenoôpetitj, religion expressed and reconfigured place, shaping and articulating changing social and geographic boundaries. For the English settlers, their Presbyterian faith was a cornerstone of the new community that was being built. People met in cabins and homes to pray and study together, and they eventually built their first small church around 1835. Religion played a crucial role in cementing the relationships of the settler community to one another and this place. As the English community grew and stabilized, they constructed first one church and then a larger one, physical signs of their permanence and prosperity. That church building, now known as St David's United Church, marks the geographic centre of Burnt Church and remains at the heart of community life in the present.

For the Mi'kmaq, their entire world view came under threat with the encroachment of the colonial governments, who reduced and eliminated traditional means of providing for family and attempted to indoctrinate people into a Christian world view. In this context, religion was a tool of the colonial project. In the earliest times of contact, religion was a part of the negotiation of relationships between the Mi'kmaq and the French. The Mi'kmaq adopted Catholicism, and rejected Protestantism in favour of Catholicism even in the long absence of Catholic missionaries (Upton 1979, chap. 11). Later, as the colonial project changed from an attempt at relationship to an effort to transform indigenous places into Canadian places, religion was central to the processes of enfranchisement and assimilation. Traditional cultural practices such as dances, festivals, and ceremonies – including gift giving, wearing of traditional dress without the permission of the Indian agent, and traditional funerary practices – were outlawed by the colonial government with the local support of the Roman Catholic Church (Paul 2000, 277). The power of the church grew as efforts to erase the Mi'kmaw nature of this place increased.

In the late twentieth century, traditional practices and languages remained under great threat. Because so many Mi'kmaw practices had been forbidden for so long, even when these practices were no longer illegal many people were initially fearful to take them up again or were uncertain as to the proper ways to enact traditional teachings. Though most adults had been raised speaking Mi'kmaq as their first language, many of the children of the community spoke only English, the language of popular culture. A traditionalist movement began to gain ground in the community of Esgenoôpetitj, encouraging people to educate their children in their own language and reclaim their traditional practices and world views. Powwows began to be held in the community at the end of each summer. Traditionalists sought to reinforce Mi'kmaw culture and identity in this place and found an avenue for their hopes in the *Marshall* decision's affirmation of their rights.

As the dispute subsided, both traditional and Catholic religious traditions remain strong and important in Burnt Church. In practice, in Burnt Church both Christianity and traditional religion are now Mi'kmaw religions, playing crucial roles in the dispute and its aftermath. Though St Anne's parish has a smaller group of regulars, the church is still full at Christmas. And though few participate in all the traditions of their ancestors, everyone in the community now comes to the powwow each August. Much as the two names of the First Nation reflect the joint identity of the people as Mi'kmaq and Canadian, religious life in the communities reflects traditional and Christian world views. These religious approaches are not necessarily in conflict with one another among the people; within families and for individuals, both can be important sources of guidance, inspiration, and support. People turn to religion as they negotiate the outcome of the dispute and what it means for themselves and their community, as will be discussed in chapter 4. Religion remains one of the ways relationship to place continues to unfold.

Contested Places

If one is trying to understand the dispute in Burnt Church, as we are, one of the most important things to appreciate is that this is a contested place and a colonized one. In *Domicide: The Global Destruction of Home*, Porteous and Smith argue that the destruction of home is a "special trauma," because the victims themselves survive, though purposely and forcibly removed from the places in which their lives have meaning and definition (2001). Edward Casey points out that "to say, 'I have no place to go' is to admit to a desperate circumstance" (1993, xii). Perhaps the dispute was, at a very real level, an expression of desperate circumstances. The people of both Burnt Churches are facing displacement, to the extent that their being-at-home and retaining a sense of home as safe haven have been disrupted, both historically and through the events of the dispute.

In Burnt Church/Esgenoôpetitj, the very identity of the people and their ties to their home have been under threat since contact. Mi'kmaw alliances with the French, first cemented by missionaries, and their later treaties with the British did not alleviate the threat, violence, and loss that came with the arrival of the Europeans in Mi'kmaw territories. Mi'kmaw people died in huge numbers from diseases brought by the colonists, and over time the ability of people to earn their livelihoods in their traditional ways and on their traditional territories was lost as settlers encroached and the colonial governments coalesced. As we have seen, loss of place has a profound effect on a people. This loss of place, along with the ongoing threat of further loss, has deeply shaped the Mi'kmaq of Esgenoôpetitj. Colonization has diminished the lands of the Mi'kmaq. At the same time, the people's experience of displacement has become a part of the place. As the place to which the Mi'kmaq retain rightful relationship becomes smaller and smaller, the precariousness of this relationship becomes a feature of the relationship itself. Sense of place in Esgenoôpetitj has become something that encompasses loss and prepares to defend and advocate against the possibility of further loss. The Burnt Church First Nation is a contested place, where "what is there" echoes what has been lost, and where "what is there" is held onto and defended all the more closely because of it.

The long experience and ongoing possibility of Mi'kmaq displacement has become inscribed in this place, and people dig in to that small spot that remains safely theirs – the reservation. bell hooks describes how, for blacks in the American South, the turn to the domestic place replaced and healed the wounds of lost place and homeland (in Casey 1993). In Casey's analysis, he quotes hooks's preference of "'the segregated blackness of our community' over the neighbourhood of whites, whose porches, even when empty or vacant ... seemed to say 'danger,' 'you do not belong here,' 'you are not safe.'" Casey suggests that "the protective posture of staying put allowed the home-place to be what hooks calls a site of resistance vis-à-vis the surrounding society in general and white racism in particular" (Casey 1993, 301). In Burnt Church, the small boundaries drawn around the community by the creation of a "reservation" mark what has been taken, and also reinforce the safe "home place." People told me stories about childhoods in which native children and their mothers would hide when white people visited the reserve, staying safe. In the English community, one person recalled a time when native people who crossed the boundary into the white side of the communities did not walk on the road, but instead descended to the shore and walked along the water's edge to their destination. Though many Mi'kmaq believe that they retain the rights to relationship with all their traditional lands, inhabiting or even visiting them is not necessarily safe. Leaving the community is safest and most often done in groups. "Staying put," as hooks put it, is a way of resistance, of protecting people and place, of reinforcing a Mi'kmaq sense of place and self in the face of all that has been lost. Loss of place is important not only as absence, but because it shapes people's relationship to that which remains.

The people of the English community have been settled on their lands for more than two centuries. Though life was challenging and difficult, as were many rural lives in Canada, for much of that time the people of the English community lived in a successful and stable community, relative to their neighbours. They, like other Canadians, reaped the putative benefits of the colonial process for generations. It wasn't until the late 1960s, for example, that native people in Canada were recognized as citizens and given the right to vote. Until at least the 1960s, the English-speaking citizens of New Brunswick tended to be much more socially, politically, and economically powerful than their French-speaking counterparts. In the 1980s, a cultural revival began in the Burnt Church First Nation, and with the Supreme Court's *Sparrow* decision on traditional and ceremonial fisheries in the early 1990s, the Mi'kmaq began their small fall fishery. With the eruption of the dispute, people in the English community had to confront the message that their presence was not entirely welcome to some of their neighbours; that they were being held responsible for the hardships of colonization and the legacies of their ancestors; that some of their neighbours thought they should "go back to where they came from."

Colonization is a thing of the past to many in the English community – a concern of history. But the dispute has torn open the worn fabric of relationship in the Burnt Churches and forced what was hidden into the light. Many in the English community feel that it is unfair that they are held accountable for the actions of their ancestors. Some argue that everyone has burdens in their past – poverty, violence – and that it is up to each individual to rise beyond their own history rather than dwell upon it. The contested, colonial nature of this place is not something that can be personalized or historicized away. It is a complex and profoundly difficult social and political reality, one that is not only the burden of the colonized, but of the colonizer and their descendants as well.

For everyone in English Burnt Church, the underlying experience is that *this* is their place and no other. There is no ancestral homeland that would welcome them back with open arms, no other place to go. Casey describes this moment when we "confront the imminent possibility of there being no place to go" as "place-panic" (1993, ix), and reminds us that "our innermost sense of personal identity (and not only our overt, public character) deeply reflects our implacement. It follows that threats to our implacement are also threats to our entire sense of well-being" (307). For people in the English Burnt Church, place-panic comes *before* they have no place; it comes when it is suggested that *their* place, the centre and root of village life and culture, is not legitimately theirs and may someday be lost to them. Place-panic comes when the awful possibility of loss inserts itself into life. The dispute also suggests the possibility of loss of place not through outright disinheritance but through loss of livelihood. If the lobster stocks are over-fished, then there is no way for many to earn a stable livelihood and remain in Burnt Church, and so the village will die. The threat to place in this sense is not direct but indirect, and for the residents of English Burnt Church, this is the more realistic, and more frightening possibility.[11] It was this threat that impelled them to action in the dispute – that the centre of their tie to place, fishing, would be lost to them for good.

Casey suggests that displacement is more complex than the "simple" severance of the connection between people and place. He argues that violence to people is often *also* violence to places. "Everywhere we turn we find place at issue in the alienation and violence from which human beings have suffered so devastatingly in modern times. More often than we realize, the alienation is *from* (a given) place and the violence has been done *to* some place, not only to people in places" (1993, xiv). The impact of violence done to both people and places is multiplied for people in a particular place, as they cope with the violence done to themselves and also to their place. The violence of colonization, for example, was done not only to the people of Esgenoôpetitj, but to their lands as well, as they were parcelled out and turned to other uses and forms than the ones in which they had initially taken shape. But this is not the only crucial example of violence done to *place* in Burnt Church. What if we looked at the problem of over-fishing, stock collapse, and the mismanagement of the East Coast fisheries as a continuation of violence done to *place*? What if we understood the effects of global environmental change on Miramichi Bay as ongoing violence done to *place*? Violence to place in Burnt Church is not simply a historical problem, it is also a contemporary problem. In communities where settlers were (are?) the agents of colonization's arrival and impact on indigenous people, residents of both communities are now also threatened by the collapse of the natural systems upon which they rely to sustain their lives and livelihoods.

The continued existence of the small English community relies upon the continued existence of the inshore fishery as a primary employer, and of the forests to supplement their incomes. The 1300 residents of the First Nation also rely on the woods and the waters for their livelihoods to some degree, as they always have. Members of each community feel that the DFO is not doing a successful job of managing the fishery; management policies often appear to them to accommodate political interests ahead of scientific ones. And to people in each Burnt Church, it seems that the good of the fishery is being (or has been) made a political sacrifice by the government, the police, and the courts in order to appease the other community or the will of Canadians writ large. And so violence to place continues, its roots (and hopes for resolution) lying as much outside these communities as in them. In chapter 5 the intricacies of this relationship between conservation, place, and livelihood will be explored further. Conservation became an important language through which people expressed their concerns for sovereignty and agency, as it was a framework that outside groups such as the government were willing to recognize and engage. This chapter's discussion demonstrates that the roots of displacement in each of the Burnt Church communities lie not only in violence between peoples but in the ongoing violence to resources, land, and livelihood that disrupts and threatens relationship to place.

In the philosophical literature on place, it is often argued that displacement is a general social characteristic of the modern and postmodern worlds. Casey argues that all modern places are essentially the same, and that desires for interchangeable places and rapid movement have displaced non-natives in the Americas just as loss of land has displaced natives (1993, 38–9). For Mugerauer, homelessness, or "displacement from both the problematic and sustaining relationships with the natural environment, community and ethos, and the sacred," characterizes the modern condition (1994, 153). It is true that displacement is a critical problem, related to globalization, the importance of speed and border permeability in the movement of transnational capital, and the shift of people from rural livelihood to the urban job market. In a very real sense, displacement is also the direct aftermath of colonization. In Canada, for example, colonial governments and settlers believed for centuries that indigenous people were a dying race, whose departure (or, possibly, assimilation) would leave the land free for Canadians to settle (see, for example, King 2003). In the present day, revitalized Aboriginal communities are working to reclaim what is rightfully theirs. This project forces settler society across Canada to confront (or repress) the question of its legitimacy and the possibility of "having no place to go," much as people in English Burnt Church confront the same problem. On the one hand, loss of place in Burnt Church is a dimension of this larger displacement created by colonialism and globalization – a very particular instance of the phenomenon of our dissociation from the very grounds of experience. On the other hand, focusing on displacement in Burnt Church as an instance of this general social dislocation glosses over aspects of the fundamental character of this place. Paradoxically, the people of these two communities inhabit places constructed around the threat of displacement, and yet they remain profoundly in place and profoundly local, tied to and inhabiting their places fiercely.

The two communities of Burnt Church are both threatened by displacement and deeply in place. For those of us living in urban North America (most of the North American population) the significance of this implacement can be easily overlooked – but the peoples of Burnt Church are not (yet) displaced as we are. Their implacement is precisely what makes the threats of resource collapse and land loss so challenging. Perhaps it is most appropriate to characterize these communities as inhabiting *contested* places, places where belonging and power are in constant confrontation and negotiation. Viewing these communities only as displaced overlooks the very reason why displacement led to overt violence in 1999: the people of the Burnt Churches know their places as home, and neither community will allow that bond to be broken easily, or perhaps at all. In both communities, people are settled on sections of land that were their ancestors', each section of each community traditionally belonging to one family. You can understand who people are – deduce their family names and all that comes with names – simply by knowing which "place" is theirs.[12]

In cities, highly mobile populations made up in part of the descendants of farmers and villagers struggle to make places for themselves in the homogenizing urban environment (see Mugerauer 1994, Stefanovic 2000). In Burnt Church/Esgenoôpetitj, many people still inhabit lives and livelihoods that depend directly on the natural resources of the lands and waters. It is tempting to think of rural communities as dying, or to romanticize them as the lost communities where we knew our place. Certainly, absolutely, Burnt Church is buffeted by the forces of modern culture, of corporations and governments and other bureaucracies, the forces that drive displacement and draw us to cities. And yet the effects of these forces, and the ways they are resisted and accommodated, remain in part unique and local. Our emphasis on the challenges of displacement in Burnt Church must proceed from and be embedded in an understanding of the profoundly placed nature of these rural communities. Burnt Church is a *contested* place. In Burnt Church, displacement is being struggled with, being fought, being challenged. In Burnt Church, displacement is a highly important factor in community life and place and has been for centuries. And in Burnt Church, people remain profoundly in place, negotiating and creating history, community, religion, landscape – inhabiting place.

4 Seeking Justice: Rights and Religion in the Dispute

The years of the dispute had a huge impact on the everyday life of the residents of Burnt Church/Esgenoôpetitj. There was tremendous upheaval in the day-to-day routes and routines of the two communities, and people found themselves having to confront violence or conflict as a regular occurrence. Over time, this had a significant effect on people's identities and world views. In the previous chapter, I considered the development of Burnt Church/Esgenoôpetitj and the English village of Burnt Church as contested places, over many generations. In this chapter, the more immediate effects of the dispute on sense of place in these communities (and of these communities' sense of place on the dispute) become the subject of discussion.

In the First Nation, people's belief that they were standing up for indigenous rights and sovereignty in their traditional territory gave them the strength and clarity to persevere. In the English community, residents began to raise questions about their own governments' inability to protect them from the ongoing conflict. In both communities, these issues remained critically important after the dispute had subsided, when things were supposed to have returned to "normal." People continue to negotiate and articulate the importance of rights, sovereignty, and nationalism as key aspects of personal practice and community life and as important parts of their senses of place. Rights, sovereignty, and nationalism are not only critical issues raised by the *Marshall* decision itself, but fundamental problems for the people and places of Burnt Church, both during the dispute and in its aftermath.

While concern for rights motivated and engaged members of both the First Nation and the English village of Burnt Church in the dispute, how these two groups understand and value rights differs significantly, as their conceptions of rights have developed within different cultural and religious paradigms. These differing views underpinned people's actions during the dispute, and how they characterized their experiences afterwards. Before exploring the significance of rights and the related concepts of justice, sovereignty, and nationalism to the dispute, it is important to attend to the differences in how these are understood in each community.

Within the Mi'kmaw community of Esgenoôpetitj, people engaged with the dispute, as one community member said, in order to "take a stand against our rights getting smaller and smaller and smaller, and [soon] they'll be worthless." Many people in Esgenoôpetitj understand rights within the context of their native view of the world, embedded in the relationships of community, family, and nation and enshrined in their agreements with the Canadian government through the treaties. For some, "rights" is not in itself an Aboriginal concept, but a Western idea encountered through colonization: "We never had to fight our Clan Mothers and Sachems for our inherent right to be and to exist" (gkisedtanamoogk 2007). In this sense, "rights" talk in the Mi'kmaw community also represents an effort by indigenous people to express their concerns in the framework of a non-native language and culture. "Like much of the english language usage in NDN [Indian] Country, the way and understanding of this usage [rights] is quite a bit different than the standard mainstream, because the usage for NDN People is culturally based" (gkisedtanamoogk 2007).

Vine Deloria argues that there is a great gulf between Western thinking about religion and culture and native cultural and religious practice, and that this difference is at its core "the difference between individual conscience and commitment (Western) and communal tradition (Indian)" (2003, 274). He suggests that the foundation of native social order is the sanctity of the individual *and* the group, and that Aboriginal religious and cultural meaning and identity emerges from the fundamental relationships of family, community, and nation. In her work on Mi'kmaw legal consciousness, "Koqqwaja'ltimk," Jane McMillan recognizes that justice for the Mi'kmaq is family based (2002, 356). Crime is understood as "an offense against relationships" (363), and justice is created by forgiveness and healing (363). Canadian law distinguishes between Aboriginal and treaty rights, both of which are recognized by the constitution. Aboriginal rights originate with the recognition that Aboriginal peoples were the original occupants of the land and have particular rights due to this prior occupation (Wicken 2002, 6). Treaty rights are based specifically in the agreements that indigenous peoples signed with the British and Canadian governments over many historical periods.

In the dispute, rights became a critical way in which the Mi'kmaq of Esgenoôpetitj expressed their concerns for justice. Indigenous "rights" are understood to be a part of the responsibilities and obligations embedded in communal relationship, belonging to the community as well as to individual Aboriginal persons. The people of Burnt Church also argue that their rights are enshrined in the treaties made between the Mi'kmaq and the early British colonists (treaties that were the basis of the *Marshall* decision). On 10 March 1760, Mi'kmaw representatives signed a Treaty of Peace and Friendship with the British lieutenant governor, one negotiated by three Mi'kmaw representatives and later ratified in many Mi'kmaw villages, including those along Miramichi (Wicken 2002, 201–2). The 1760/1 treaties are related to the earlier 1725/6 treaties signed between the Mi'kmaq, Maliseet, and Passamaquoddy and the British colonies of Massachusetts, New Hampshire, and Nova Scotia at Annapolis Royal. These treaties outline relationships between the Mi'kmaq and British, guaranteeing terms of peace and trade between the two nations.[1] In Esgenoôpetitj, people look to both these treaties as documents that outline their collective rights as Mi'kmaw people – treaty rights that were recognized by the Supreme Court in *Marshall*.

For most in Esgenoôpetitj, the ability of the community to enact its rights as outlined in the treaties would represent the realization of justice for the people. Justice, in this view, cannot be realized while members of the community enacting its fishing rights are being pursued and prosecuted for their actions by the Canadian government – and persecuted by their neighbours. For a smaller group within the First Nation, the just realization of their treaty rights requires not only the recognition of these by the Canadian government, but the recognition of the Mi'kmaq as a sovereign nation by the Canadian government. Sovereigntists argue that as the treaties were signed between nations, the Mi'kmaq nation retains the right to govern its own activities, such as fishing or timber harvesting, without the oversight of the Canadian government. Regardless of their position on sovereignty, all whom I met in Esgenoôpetitj agree on the significance of rights, as a communal good, to the dispute.

In the English community of Burnt Church, rights also became an important concern during the dispute. When the community held a protest against the native fishery, on the wharf, some carried protest signs that read, "What about our rights?" Some residents argued that developing a separate set of rules for a native commercial fishery was wrong: "If they've got a law, everybody's got to abide by it, by my way of thinking. You can't have different laws for different people depending on your race, colour and creed." The concept of rights at work in the English community supposes that rights reside with individual persons, not with groups or communities, and that rights need to be equally recognized and distributed among these individuals. This notion of rights as inherent in individuals reflects the common Western (nonnative) understanding, as is outlined in the Canadian Charter of Rights and Freedoms, which ascribes rights to individual citizens (Department of Justice, Canada 1982), or in the Universal Declaration on Human Rights, which recognizes rights as residing in individual persons (United Nations 1948). Within the English village, this understanding of rights as individual leads to the view that justice is the equal treatment of every individual, according to the same rules. The presence of a native fishery, even when recognized by the Canadian courts in *Marshall* and in *Sparrow*, did not uphold an understanding of rights or justice that the people of the English community affirmed. This led to deep divisions between local residents and the governments and government agencies that they had expected to uphold "one law for all."

Within environmentalist circles, the question of the rights of ecosystems themselves has become important as the environmental movement has developed. Many who've heard me speak about my research in Burnt Church/Esgenoôpetitj are initially tempted to draw connections to deep ecology. Deep ecologists argue that non-human nature has inherent rights, particularly at the systems level, and that recognition of these rights is an important first step in enacting an ecological cosmology (see, for example, Devall and Sessions 1985, Macy 1995). Such concerns were simply not on the table among those with whom I spoke in Burnt Church, or in any of the related literature. Concern for the lobster was never separate from concern for the lobster fishery; in conversation, rights were ascribed to people and never to crustaceans (or other non-human creatures). Local understandings of environmental concerns encompass human and non-human nature, as will be explored in detail in chapter 5, but rights were never ascribed to ecosystems as such.

Life in Burnt Church during the Dispute

In addition to understanding the underlying principles which motivated people's perseverance in the dispute, it is equally important to attend to people's lived experiences of the dispute. These experiences, and the ways that they are characterized by the people who lived them, give us further insight into rights, sovereignty, and nationalism in this context. Understanding the ways the dispute shapes and is shaped by sense of place requires turning again to the events of the dispute itself.

In the First Nation/Esgenoôpetitj, when people were fishing, the attention of the whole community focused on the shore. The fishing was begun by some of the women of the community; but when the violence from the government and their neighbours escalated, the men took the lead on the waters, supported by the broader community along the shore. Some stories describe how people "created their own justice" from within the community because they didn't believe that they were experiencing justice within the Canadian system. This "thirst for justice" is part of what motivated many to take to the waters on behalf of their community during the dispute. One community story tells how some of the non-native (Acadian) commercial fishers were chasing native fishers on the waters, taunting them with caricatured enactments of native culture – face paint, made-up chants, and tomahawk dancing – and threatening to harm them and their fishing gear. Eventually, one boat ran aground – the men on it had been drinking – and the men and their boat were captured by people from the native community. The native boats took action against the protesters, eventually turning the men over to the RCMP and burning their boat. One native fisher characterized this action as sovereign action – the people enacting their own justice: "And the thing with that was when [the community] did it, you know, it was a great relief in their hearts, because when we did that, they [Acadian fishers] never came in our waters again. Because sometimes you have to have your own justice. Because the justice that we were gettin' here wasn't too good." The creation of justice for and by the people is an important dimension of sovereignty for many in Ensgenoôpetitj. The people's anger is channelled, and the government challenged, by the creation of Mi'kmaw justice. The creation of this justice brought "a great relief in their hearts."

The residents of Esgenoôpetitj were closely monitored by RCMP officers on the roads and in their homes; the community was under helicopter surveillance; it was openly acknowledged that phones were being monitored by government officials.[2] Whenever there was a raid by the Department of Fisheries and Oceans (DFO) or a confrontation on the waters, native people would drive their cars through the reserve, honking their horns. Everyone would recognize this as a signal to go to

Figure 5. After a provocative demonstration against the native fishery on Miramichi Bay, one non-native fishing boat ran aground. The boat was burned, and its occupants turned over to the RCMP by the Mi'kmaq. In this picture, the remains of the burned-out non-native fishing boat rest in the waters of Miramichi Bay, with St Anne's Parish Church in the background. (Monday morning, 17 Sept. 2001. CP Photo/Jacques Boissinot)

the shore and witness their fishers being confronted by armed DFO and RCMP officers, who were destroying or confiscating their traps. One couple with whom I spoke described the effects that watching DFO raids had on them:

> It was a constant attack on our dignity, that's what it was. I felt so helpless – and you couldn't do nothing! Ah! You'd see them all out there, and you'd know they were cutting [traps]. And even though I don't fish or anything like that, they weren't just doing it to individuals; they were doing it to the whole reserve, the whole people. And not for what we were doing, but for what we were, as natives.
> ... Each time they came close to a boat, they were taunting, and they were telling us, "We'll get you!" "Watch your back, we'll get you!" ... They had most of the people's profiles. They knew each and every one of the people that went out and protected the traps, they knew their background ... We had files with the Security Services. (Dalton and Cindy)

The threat to the traps and to the men on the boats was not felt as simply a threat to those individual persons. It was something that was sensed across the whole community. The women and men who remained on the shore watching the conflict unfold on the waters felt the threat to their brothers and uncles and sons very deeply. They also felt the responsibility they had as a community to keep all of their members safe, especially those who were risking their lives in the confrontations on the waters. A younger woman in the community described the fear that the conflict created in her mother-in-law, an elder:

> My mother-in-law, that was hard for her, to see all her boys out there ... I remember my husband went into a battle, and when it was finally settled down, his mom walked up there. It was right in front of her house, you ... can see everything right in front of there. And she was holding onto his hand ... and then she started crying, "I thought you were going to get killed out there." He said, "I'm all right Mom, I'm okay." And then I could see him crying too, because he feels the pain for his mom, eh? So then his mom finally kind of turned around, and gave a big look, and say, "I know what you have to do, son. I believe in what you are doing, don't stop." (Audrey)

People on the shores were galvanized by what they witnessed on the waters, and tried to find many ways to support the actions of the fishers. One of the most important actions was always that of being present on the shore, to witness and support the work of the fishers and the Rangers (band-appointed fisheries officers) on the waters. The work of keeping the community going was also important, taking care of children, feeding visitors, and boosting the spirits of those on the front lines:

> During the day I would stay home and make sure the kids were safe – I wouldn't even let them go out in the yard anymore ... We were just so scared ... I just mostly stayed home and took care of the kids. And supported my husband, because there was a lot of times he was discouraged, and didn't know what to do, and said "I don't want to go out there anymore ..." But then we look around, and I told him that "I believe what you're doing is right. And they need you." (Audrey)

Another woman described the work that was done to host visitors, and how it brought the community together:

> Strangers would come by, delivering 50 lb bags of potatoes, boxes of food, and they would drop them off at the band office, because the band office was the kitchen of the community. That's where all the cooking was going on. There was people designated to do the cooking every day, to feed the Warriors, to feed the guests. And people all just pitched in, and they worked side by side. Much as they were fighting [each other] before [the dispute], they were fighting together [now]. [*laughs*] (Barb)

The experience of the dispute drew the Mi'kmaw community together to work for a common cause. Many people with whom I spoke noted how community members overcame differences, addictions, and despair as they cooperated to support the fishery.

Meeting practical needs of the community was not the only way that those on the shores supported the action of the dispute. They erected barricades on local roads and highways in response to raids on the water, which they operated in concert with the Warriors who were in the community. At the barricades in Burnt Church, anyone attempting to pass would be confronted. While local vehicles were usually let through, commercial vehicles were forced to take long detours around the community.

Sometimes the people standing on shore picked up rocks and hurled them at government boats in an attempt to protect their fishers:

So sometimes we would be out there, and there would be a conflict taking place.

They'd tell the young men, "Don't engage in anything with them, just go out there and fish." But when [the government] started ramming the boats, they said "That's it!" And they started collecting all these rocks – "Come back! Come get your rocks!" And the people like, there were grandmothers and grandfathers and little children, they were filling up buckets with rocks … because nobody had any guns, and they had the guns. Not the Warriors or the fishermen, we didn't have any guns, but the DFO did.[3] (Barb)

The violence of the dispute had a huge impact on the community of Burnt Church. People lived under a very real and present threat for many years. They witnessed and experienced violent raids and arrests, including beatings and capsizings, and the constant looming presence of the government's agents, whether confronted by them in full riot gear with weapons pointed or through their surveillance and monitoring of community life. The resolve of the community during these extreme circumstances was deep. The roots of this resolve lie, in part, in the significance of treaty rights and sovereignty to the members of the Burnt Church First Nation, and in their determination to win decision-making power over their own place as they believed the *Marshall* decision finally recognized.

Though the English and Mi'kmaw communities of Burnt Church are largely separate, they live in the same space and landscape, drive on the same roads and rely on the same resources. The prolonged conflict of the dispute was primarily between the Mi'kmaq and the federal government, but it did not begin that way. In the very first days of the post-*Marshall* native fishery, in the fall of 1999, the federal government was clearly unprepared (Coates 2000, Doyle-Bedwell and Cohen 2001). They had not anticipated that the court would rule as it did, and were not engaged in any conversation with Mi'kmaw people about the fishery. In the English village of Burnt Church, after *Marshall*, fishers and their families watched unlicensed native fishers fishing lobster out of season and then selling their catch to local buyers and processors without any action from the DFO. As the native fishery continued, the people of the English community were motivated to organize themselves into the demonstration that took place on 3 October 1999. Burnt Church fishers took their boats out on Miramichi Bay, where they were joined

by many others from the surrounding Acadian and English communities. Their families and other local residents marched to the wharf, carrying Canadian flags and protest placards. When some fishers on the
Bay began to cut native traps, the protest famously erupted into violence between the two communities.[4] The RCMP eventually escorted
the non-native protestors off the wharf. Non-native trucks were burned,
boats and lives threatened, and a violent altercation broke out. At the
end of the afternoon, Mi'kmaw residents occupied the wharf. The conflict between the two communities had broken wide open.

In his examination of the effects of the *Marshall* decision in Maritime
Canada, Coates argues that the non-native people of the Maritimes
simply had no idea of the depth of pain and dissatisfaction with the
status quo that existed in Aboriginal communities (2000). In Burnt
Church, English residents were suddenly confronted, in their neighbours' reaction to their protest, with this usually hidden anger. The resulting mobilization of native people, and their occupation of the lands
of both communities, had a lasting impact on the English community.
One English resident described how these effects began the day after
the protest and the native occupation of the wharf:

> On October 4, the day after all this took place, when the other natives
> started moving into the area, they were dressed in camouflage outfits,
> some of them had masks on. They set up three different teepees – two of
> them out on the wharf, and then there was one down here on the shore …
> And of course they were going night and day at that point in time, these
> people that had come in from other areas. Of course, that bothered the
> community quite a bit. All of a sudden a quiet little place like this is dis-
> rupted with people tearing around 24 hours a day … So we didn't really
> know who was around or what they might be up to. Whether they were
> gonna set your house on fire or what … It was a very difficult period. Was
> just like the community was under siege, for a period of time. (Paul)

This feeling of siege was very acute for the residents of the English community. They were told by the RCMP to be concerned about their safety,
that their Mi'kmaw neighbours might come after them. Some were told
that their names were "on a list of people the Indians were looking for."
Many were told that they should leave their homes and the community
and find another place to stay since the RCMP could not guarantee
their safety. Every resident (save one) with whom I spoke about this
time, whether informally or in interviews, said that in the days

immediately following the protest and occupation, someone from their household sat up at night in their living rooms, trying to protect their families and properties from the threat that the Mi'kmaq posed. Some followed the advice of the RCMP and left, or sent vulnerable family members away temporarily.

> I sent my wife to go upriver ... cause I knew what was going to take place here. And my young lad was in college, and he heard about it, so he come home ... and we sat up at nights with shotguns, because we couldn't sleep. Like if cars were going right by, they'd slow right down ... at your gate, you didn't know if they were coming in or what. (Luke)

Suddenly, over the space of one weekend, the perception of life in the English community as basically safe and stable was transformed.

As the government presence in the two communities increased over time, the direct threat that the English felt from their neighbours shifted. The possibility of violence became less a direct threat to individual homes and people and more a constant presence in the public spaces of the community. The wharf, maintained by English fishers and at the centre of the English community, remained under native occupation, as did a stretch of the "English" shore.[5] The place that the English had thought of as "theirs" was under occupation by their neighbours.

> The reserve has always been a place that you don't go ... but it never really spilt out into our community ... it was "over there" ... So when they decided to take over the wharf ... and thousands of native people were there, where did they come from? Who are they? Why are they here? It was like an invasion – our wharf, of our space. (Brenda)

The barricades that were erected included one along the shore road near the wharf, within the bounds of the English community, and one at the main crossroads of the English community. This meant that some residents had to cross the barricades in order to carry out the functions of their daily lives, like picking up the mail, buying gas and groceries, visiting neighbours, or going to work. The presence of the barricades was painfully disruptive to the English residents. Their anger was compounded by the fact that some of the people involved in these protests were not local residents of Esgenoôpetitj but Warriors who had come from other communities in the region, and from as far away as British Columbia. Native protesters were an ongoing presence in the public spaces of the English community, on the wharf, on the shore, and on the roads: "They'd stand on the picnic tables down there at the shore, and they'd be having their lobster feeds, and music blaring, and loudspeakers and the whole nine yards, you know! ... and I did say to the RCMP one day ... that I feel that I'm being watched. Mostly they're just intimidating you" (Martha).[6] While the protest was, for the Mi'kmaq, a dispute over rights and sovereignty, one of its effects was to diminish the comfort and security the English residents felt in their own homes. Much of Canada was writing and talking about events in Burnt Church, yet people in the English community often felt that they were the only Canadians who actually understood the consequences of, and had to pay the price for, the *Marshall* decision. The violence of the dispute and its presence in the public space of the English community had a significant impact on people's views about native rights; the unpredictable responses of the Canadian government have led to a renegotiation of people's understandings of themselves as Canadians. The "end" of the dispute has also led to a re-examination of values and priorities in the First Nation, as people seek new tools to meet the needs of their community. As a later section of this chapter will demonstrate, in the English village people have moved to reclaim the public space of their community as Canadian rather than as occupied space, and in this way to reclaim their Canadian identity on their own terms.

Rights and Sovereignty in Esgenoôpetitj

Rights and sovereignty were common concerns that motivated the people of Esgenoôpetitj during the dispute, and their commitment to these priorities is what allowed them to maintain their fishery for so long. People in the community have a very clear and articulate understanding of their rights as Mi'kmaw people, and of how these came to be. The treaties that were the basis of the *Marshall* decision are public documents, hanging on the walls of some homes. Community discourse around the dispute highlights the importance of rights as tools to call the Canadian government and the elected (Indian Act) chief and council to account, and of sovereignty as a goal that is simultaneously political, spiritual, and geographic. For some, rights and sovereignty enable them in the ongoing challenge of re-establishing Burnt Church as a Mi'kmaw place. The outcome of the dispute, a temporary agreement in which many people feel their rights were subsumed under the regulation of the Canadian government in return for boats, training, and money, does not address sovereignty or rights in any substantive way.

The *Marshall* decision upheld important treaty rights of the Mi'kmaq, particularly around access to the fishery. It recognized the right of Mi'kmaw people, under the peace and friendship treaties, to earn a moderate livelihood by fishing. The dispute was, in a very real sense, a conflict over how this right would be exercised and under whose authority and regulation. The few existing academic writings about the dispute focus on this issue and on the importance of *Marshall* to treaty rights, as outlined in an earlier chapter. Linking legal discussions of *Marshall* with the events in Burnt Church, Doyle-Bedwell and Cohen attempt to balance insights from the Canadian legal system with those of traditional Mi'kmaw world views. They argue that "the current conflict over the Marshall decision illustrates the difficulties faced by the Mi'kmaq people and the federal government when the government fails to acknowledge the extent of its fiduciary obligation to include Aboriginal people in resource management decisions" (2001, 193). In recent years, the Canadian government has devolved many of its fiduciary responsibilities to provincial and territorial governments. This presents many challenges to native communities, as they seek a responsible relationship with a government that no longer recognizes some of its original obligations under the treaties. For some with whom I spoke, the dispute was an effort to call the federal government to account for its responsibilities to First Nations people. People were seeking a relationship between the government and the Mi'kmaw people which upheld the principles outlined in the treaties, as they understood them.

> The government, through the Indian Act and British North American Act, all these acts, puts the government in a fiduciary responsibility for the native people in Canada ... They're trying to find ways to get rid of that traditional responsibility ...
>
> Our health, medical, used to be under the federal government, they unloaded us to the province ... This is what the fishery was all about. To take a stand against our rights are getting smaller and smaller and smaller, and [soon] they'll be worthless. (Dalton and Cindy)

Concern for Mi'kmaw rights, and the Canadian government's responsibility to rights and rightful relationship, was articulated in literally every interview and conversation I had about the dispute in Esgenoôpetitj.

The *Marshall* decision affirmed the community's own understanding of their history and motivated them to act upon it, using rights as a tool to point to the gap in their relationship with the federal government.

With the signing of the interim agreement between the Indian Act chief and council and the federal government, more people in the community are now fishing, with proper gear, and this is positive for those who were concerned specifically with fishing rights.[7] Not all community members feel that the chief and council had a clear mandate to sign the agreement with the federal government. This agreement is seen as a mixed victory, since it allows for a native commercial fishery but not under native regulation and governance. Some feel that their ability to continue to agitate for self-regulated fishing rights is undermined by the presence of fisheries officers from their own community who work in cooperation with the federal government: "We've just been put in a really bad situation where we would be fighting each other" (Miigam'agan).

Many argue that the benefits that were supposed to come to the community with the signing of the agreement were not distributed fairly to all, accruing instead to a few specific people aligned with the chief and council and to the non-native fishers in surrounding communities who were paid to carry out training for natives entering the fishery. But there are also some situations where the distribution of fishing boats to families in the community has made a difference to incomes and well-being. Criticism of the chief and council and of the agreement, and strong feelings of despair and discouragement, dominate the aftermath of the dispute for many; but for others, some hope exists that the boats now in the community will continue to make some kind of small difference to the lives and incomes of people who can fish, even if this is not accompanied by a change in government or government relationships.

For those in Burnt Church/Esgenoôpetitj who are or have been proud of their identity as Canadians, the federal government's position in the dispute was a critical problem. Why was it that their own government refused to address their concerns? Why didn't their elected representatives take their concerns seriously? And what are they to do now that their own government has shown it is violently against them?

> They [our governments] don't see us as Canadians. They're looking at us as people that they have to put up with, because they moved to this country and they don't know how to deal with us. But they know that they have to deal with the non-native people. ... They think that the

non-natives have more at stake because they're taxpayers. And because we're not taxpayers we don't receive … the same respect. (Barb)

The experience of the dispute, in which Mi'kmaw treaty rights as outlined by the contemporary Canadian courts were not recognized by the Canadian government, reinforces the "outsider" status of the Mi'kmaq to Canada, especially for the people of Burnt Church/ Esgenoôpetitj. In their own community, many people do not feel able to enact and regulate their rights in their own terms, as they had hoped and fought for. For many Mi'kmaq with whom I spoke, the Burnt Church First Nation remains a colonized place, where the will of the federal government disregards local indigenous interests.

Sovereigntists in Esgenoôpetitj see themselves as members of the Mi'kmaq Nation, and not Canadians at all. This more radical position, held by many of those who were community leaders during the dispute, is also grounded in the treaties. For sovereigntists, it is precisely because the Mi'kmaw people never ceded any of their lands or voluntarily extinguished any of their rights that the legitimacy of traditional Mi'kmaw government over Mi'kma'ki (Mi'kmaw territory) must be recognized. The dispute was seen to be not about fish, or only about fishing rights, but about the right of the Mi'kmaw people to govern themselves and Canada's systematic and historic denial of that right. Sovereigntist resistance was not only against the Canadian government, but against the structures it imposes upon the community, including the elected chief and council.

The chief and council are elected and govern according to the regulations of the Indian Act, a Canadian colonial framework. The Mi'kmaq Grand Council (Sante' Mawi'omi wjit Mi'kmaq or Holy Gathering of the Mi'kmaw) is a traditional form of government that has existed in Mi'kma'ki for centuries. Its existence is documented in the writings of some eighteenth-century colonists, and according to Mi'kmaw oral tradition it dates to pre-contact times (Wicken 2002, McMillan 2002). Since the eighteenth century, the Grand Council has functioned as a Mi'kmaw institution and been shaped by Mi'kmaw engagement with Catholicism and with British and Canadian governments.[8] Grand Council leaders are chosen through a community process that has nothing to do with the Canadian government and goes unrecognized by it. The chief and council system is understood by sovereigntists to be responsive to and an imposition of the Canadian government's colonial efforts.

Sovereigntists advocate for a traditional government arising out of Mi'kmaw culture and the will of the people.

As Lloyd Augustine (Kwegsi), the keptin or traditional chief explains it, the Grand Council is a form of Mi'kmaw government that has its origins in Mi'kmaw civilization long before colonization. Lloyd sees the Grand Council as a key source of traditional teachings on Mi'kmaw governance and politics. For other sovereigntists, such as gkised-tanamoogk, the accommodation of colonial interests by the Grand Council – he cites its move towards a more patriarchal and Catholic structure after colonization as an example – means that its contemporary form has diverged significantly from traditional understandings. Although precisely how "traditional" the Grand Council remains is contested, the importance of a traditional alternative to the chief and council system is a critical idea among all sovereigntists at Esgenoô-petitj, regardless of their view of the Sante' Mawi'omi. For some, the continued existence of the Grand Council serves as a reminder that such alternatives are possible and as a connection to historic structures of Mi'kmaw self-government.

During the dispute, the keptin and other non–Indian Act community leaders were important figures in the work of the community. As the pressure from the Canadian government mounted over the years of the dispute, the power of the elected chief and council rose. In the band council elections of 2001, a sovereigntist leader at the forefront of the dispute ran against the elected chief and lost. With the signing of the interim agreement, the traditional and community-based leaders who had eclipsed the Indian Act chief during the dispute have faded from the forefront of community life. While the sovereigntist activists feel good about the stand they took, the agreement made with the government does not recognize their goals of sovereignty, justice, and Mi'kmaw management of Mi'kmaw resources.

The "loss" of the dispute was profoundly challenging for those most radically committed to sovereignty. As the Indian Act chief and council reconsolidated their power in the community, those who gave leadership during the years of the dispute were shut out of the economic life of the community in many ways: "Most of the people that were involved, were basically used and then dropped ... People that had jobs at that time, still have jobs now. But the people that didn't have jobs, and they were in the forefront of the dispute, they have [no hope]" (Cindy). People who worked full time in the dispute for the needs of the community often found themselves without work or support in the months after the dispute subsided.

At the same time, as people in the forefront of the dispute, they had to cope with the aftermath of the tremendous stress and trauma from the conflict and from the public scrutiny and vulnerability that they experienced. Relationships between some of these people cracked and broke. Some left the community to find work and support their families, or to heal, or to continue their commitment to activism for Aboriginal people. And for most, the "end" of the dispute was a profound challenge to their hopes for their community and for the future of their people. They believe in the importance of what they did, and its significance for future generations, but have little hope that they will see change in their lifetimes. For these people, the signing of the interim agreement reinforced the marginalization of Esgenoôpetitj as a colonized place, and again marginalized those within the community – moderates and sovereigntists alike – concerned with larger questions of rights and sovereignty.

The perception that the dispute was lost when the agreement was signed, and with it many hopes for rights and sovereignty, as well as a sense that the community could be an agent of reform and change, precipitated shifts and changes in people's world views and values. During the years of 2004 and 2005, people were quietly trying to make sense of all they had experienced, and hoped for, as they got on with the other challenges of life. People worked through their experiences in myriad ways. Among those I knew and spoke with, responses included a deepening of engagement with traditional spirituality and practice, a shift in focus towards community and family well-being, and, for some, a turn to charismatic Christianity.[9] These overlapping approaches are directly tied to the concerns for rights and sovereignty that were acted on in the dispute, and demonstrate how people are trying to negotiate their concerns after the end of the dispute. There are not firm boundaries between the approaches I describe below; individuals and families may be doing one or two or all three of these things.

Concern for sovereignty and traditional practice was certainly not unanimous in Burnt Church during the dispute. Most people got involved because they believed that the ability to exercise their economic and resource rights under *Marshall* would improve the lives of the people in the community. There are simply not enough jobs for the people who live in Esgenoôpetitj; there is a chronic housing shortage; addictions and depression are a common part of the experiences people related in their accounts of life on the reserve. In the view of many, fishing rights provided an opportunity for people to solve these problems on their own terms, to improve life for their families and their communities without waiting for government bureaucracies to do it for them. As people came together to take action during the dispute, many residents observed that conflicts and addictive behaviours within the community decreased. After the signing of the interim agreement, incomes increased for the families who got fishing boats, but things returned to the status quo for everyone else. The critical problems that people attempted to address through the dispute remain. Those who have energy for change focus on raising and educating the next generation, getting their children and grandchildren through high school and university. They do this believing that it is the best thing for their children, and also for the long-term hopes of the community. Those working at the community level focus on healing the community's ills, addressing issues of addiction and despair. These concerns continue to motivate many in the community who are not religious at all, as well as those from all three religious communities – traditional, Roman Catholic, and charismatic. They have not stopped believing in their rights and in the importance of them, but with the loss of the dispute they have set aside activism for rights as their preferred tool for change in order to focus on family and healing.

Traditionalists within the community, those who look to Mi'kmaw spiritual and cultural practice to shape their lives, have been working to reclaim traditional religion since the 1980s. They understand their place as traditional peoples and their relationship to one another and to the land in much the same way that Vine Deloria describes it in *God Is Red* (1994). For Deloria, the Indian traditions are communal ones, rather than a choice of individual practice and commitment as in standard liberal theories of religion. The well-being of the entire community, including the non-human relations, rests upon the ceremonies carried out by individuals and communities (83, 85). Traditional peoples have a moral responsibility for ceremony, a responsibility that extends not only to other people in their community but also to place and to the planet (85). This sense of responsibility grounds those from Esgenoôpetitj engaged in the ceremonies and in traditional cultural practices more generally. Traditionalists believe that through the ceremonies, new messages and understandings come to the people, messages specific to the community's life at that moment. In the aftermath of the dispute, Mi'kmaw traditionalists from Esgenoôpetitj turn to the ceremonies, but with a sense that their hopes will be fulfilled by future generations rather than in the immediate term. Prophecies are central sources of guidance and knowledge in traditional Mi'kmaw world views, as Marie Battiste illustrates in her account of "Nikanikinút-maqn," the Mi'kmaw teachings of pre-contact history (1997). One traditionalist at Esgenoôpetitj spoke a number of times about the importance of attending to the prophecies that had been given to the ancestors. Deloria describes these same prophecies in his writing:

> Long-standing prophecies tell us of the impious people who would come here, defy the creator, and cause the massive destruction of the planet. Many traditional people believe that we are now quite near that time. The cumulative evidence of global warming, acid rain, the disappearance of amphibians, overpopulation, and other products of civilized life certainly testify to the possibility of the prophecies being correct. (1994, 86)

The ongoing threat of fisheries collapse is interpreted by Esgenoôpetitj traditionalists as further evidence supporting the prophecies. For traditionalists, the prophecies, the challenge of life in Esgenoôpetitj, and the outcome of the dispute point together to a looming crisis. This deepens the importance of their responsibility to the ceremonies and of their work for healing.[10]

In traditional communities, knowledge is often shared in stories. As I discussed in chapter 2, stories were often the way that people chose to share their experiences of and insights into the dispute. In Esgenoô-petitj people shared in story regardless of their affiliation with traditionalist movements, and stories were also a preferred method of communicating in the English village. As an ethnographer in these communities, listening to these stories was critically important, as was gathering the stories of my own experiences. In anthropology and religious studies the richly descriptive sharing of stories from research is often referred to as "thick description," a necessary part of rigorous and reliable qualitative research. Stories are not distractions from the larger discussion of values and place; they are important ways of reflecting the lived experiences of local people, and of bringing their subtle insights into this book. As the discussion of religion develops here, these stories become important avenues for understanding the complicated interrelationship of religion, politics, values, and place.

Among the Mi'kmaw of Burnt Church/Esgenoôpetitj, many are not intentionally engaged in revitalizing traditional ways. But for some of those who are, the struggle remains, everywhere, constantly, guided by the prophecies of the ancestors and through connection to the non-human relations. Miigama'agan speaks eloquently about how being on the waters of Miramichi Bay during the dispute deepened her connection to her community and her traditions:

> People experienced a lot of what our ancestors had talked about, the visions that used to take hold. And so we started experiencing much of that, personally. Even out on the water, it was the presence of the ancestor's boats, the human relations and the non-human relations. It was so evident, that spiritual part of it all was that reconnection to the unseen life. And a better sense of total acceptance. Maybe it's not being practiced or remembered every day, but the kids are going to even make it more of a legend, you know what I mean? It's gonna be bigger than, even what we know of today ... In one sense, I think it's an experience I would never change, even though its hard ... It's a balance always, it's a lot of deep wounds, deep hurt, [but] the other side to that is liberation and spirituality, spiritual liberation – something you can't hold [on to], but you know.

The hurt of the dispute has refocused hopes on spirituality and affirmed the importance of the traditional ways for these people, though they do not expect change in their lifetimes. The work they undertook in the dispute and the ongoing importance of the ceremonies means that future generations will have the visions and the tools they need to make change, when the time comes.

Among other sovereigntist traditionalists, the loss of the dispute and the accompanying loss of hope for Mi'kmaw sovereignty over Mi'kma'ki (in this lifetime) has precipitated a turn to Pentecostal/charismatic Christianity. Similarly to the traditionalists above, these people have come to believe that restoration of the sovereignty and justice that they sought through political action will now come only through spiritual action. For them, God's sovereignty on earth will, in God's time, lead to Mi'kmaw sovereignty over Mi'kma'ki. This turn to charismatic Christianity in Burnt Church is not dissimilar to the adoption of charismatic practice in other indigenous communities. In Alaska, for example, Dombrowski has described how marginalized traditional peoples have turned to charismatic Christianity as an alternative to the state and forestry-corporation-sponsored expressions of indigeneity that are taking hold in their communities (2002). He argues convincingly that these conversions are not examples of successful colonial missionization, but are in fact expressions of anti-hegemonic and anti-Western sentiment among marginalized Alaskan natives (1072). This analysis has parallels among the indigenous Urapmin of Papua New Guinea, who, Robbins argues, have adopted Pentecostal/charismatic (P/c) Christianity as a response to their increasing marginalization after the arrival of Western-style development and culture within their region (2004a).

In Esgenoôpetitj, some Mi'kmaq have turned to P/c religion to help them negotiate their confrontation with colonialism. For them, P/c practice often goes hand-in-hand with indigenist and sovereigntist politics, and for some, with traditional religion. Much like their ancestors who engaged Catholicism, transforming it into a Mi'kmaw religion (see Henderson 1997, Upton 1979), those involved in P/c practice integrate native and charismatic practices and identities rather than trading one for the other. In Burnt Church, many who have turned to charismatic practice have done so as an expression of resistance to the current social and government order, and as a way to maintain their hopes for the sovereignty of Mi'kmaw people and governments in Mi'kma'ki, as shown in detail below.

Studying Christian cultures, especially in indigenous communities, is a challenge, as many students of cultural change tend to see Christianity as "the perennial outside force" (Barker in Robbins 2004a, 28). Anthropologists have tended to separate Christianity from what is "really important" (i.e., traditional culture), seeing the adoption of Christianity as a loss of traditional culture, as an insincere accommodation, or simply as traditional religion in disguise (30). This is especially true in the case of charismatic, Pentecostal or fundamentalist Christianity, which is so fundamentally "other" to the academy. Harding suggests that this presents a profound challenge, because "to recuperate them [charismatics] to reasonableness by showing … that they make sense in their own terms would be to lose our assurance that we make sense in ours" (in Robbins 2004a, 29). In Burnt Church, some dispute leaders told me clearly that if I was interested in their religion and world views, then I must attend to Pentecostal/charismatic Christianity.

As a newcomer in the community, I was introduced to some of the people who had provided leadership during the dispute, so that I might talk with them about my project and how it might come about. This included an introduction to Leo, who worked as head of the Rangers on the waters during the years of the dispute and makes his living today on the waters, in the woods, and in addictions counselling. During my second visit to the home of Leo and his wife, Audrey, they expressed quite an interest in my academic field, religious studies. They wanted to know if I studied the Bible. I explained my interest in religion as a social one, an interest in how belief and practice are important in the lives of people and communities, rather than a theological one, concerned with interpreting God's will. Despite this, Leo and Audrey invited me to join their Bible study group meeting that week.

I was not aware that charismatic Christianity was important to some in the community before I arrived in Esgenoôpetitj, but I found that my participation in this group helped me to understand much about how some negotiated meanings of the dispute and its aftermath. Robbins argues for the importance of understanding Christianity in cultural terms, and asking how it is that people live their lives as Christians (2004a, 31–2). This is what I was being invited to do in Burnt Church, as I participated in the charismatic Bible study group.[11] Many of its members are leaders in the community, and were leaders on land and in the waters during the dispute. The group also includes people from the Roman Catholic parish on the reserve, St Anne's, some who consider themselves traditional Mi'kmaq (including some sovereigntists), and others who have no religious affiliation. It's a place where theology, politics, and family come together, often in unspoken collision. As an outsider, I have characterized this group as charismatic because their practices during my time with them included prayers for healing, laying on of hands, and anointing with oil, along with discussion and exploration of being "slain in the spirit" (an experience understood by believers as being transported to an alternate state of consciousness by communion with God through laying on of hands) and of speaking in tongues. Charismatic is not a phrase that these people usually use to describe themselves (many prefer the term Christian), but I use it to distinguish their Christian practice from Mi'kmaw Catholicism and other possible models.

As strong, independent community leaders and members, these people have turned to charismatic practice. Why? What does it give them? How does it help them understand their own lives and experiences? Many participants in the Bible study group see their religious practices as deeply Mi'kmaw *and* deeply Christian *and* deeply charismatic. Lloyd, the keptin, is also a core member of this group. For him, Mi'kmaw Christianity is explicitly an act of resistance against the oppression of Canadian society and the Indian Act governments. In fact, Lloyd would be uncomfortable with my use of the word Christianity – he calls himself a "follower of Christ" as a way of separating his theological position from the history of Christian missionization in his community.[12]

Lloyd's life is grounded by Mi'kmaw culture, including some of the practices of traditional religion, by his participation in traditional forms of Mi'kmaw government, and by his understanding of his relationship with God through Jesus Christ.[13] He sees his "walk in this path with Christ" as a Mi'kmaw path, like the path of traditional religion, and sees both as encompassed within Mi'kmaw identity and culture: "We need to hang on to who we are as a people, we can't give that up. Paul [the Apostle] said 'When I called you, I called you to stay who you are.'" Lloyd says that religion for him is not about "playing church" or "playing tradition," but about the deep transformative power of talking with the Creator. In his view, these conversations with the Creator can be mediated through ceremony, as in traditional religious and cultural practices like smudging and sweat lodge, or directly through prayer, like the intensive prayers of charismatic practice. Lloyd argues, "Once you allow the Creator to talk to you and Jesus to become part of you and in you, then you see things in a different light."

For Lloyd, sovereignty and faith are linked in the face of the dispute. He believes that ultimate sovereignty belongs to the Creator. Justice and self-government will come to the people of Burnt Church, but in God's time. Charismatic Christianity is becoming a resource for resistance, a way to sustain hope for the healing of the people and the restoration of their sovereignty. It brings to mind one of the songs sung at the Bible study group.

> In the name of Jesus, in the name of Jesus, we have the victory! Hallelujah!
> In the name of Jesus, in the name of Jesus, demons will have to flee!
> When we stand in the name of Jesus, tell me, who can stand against us?
> When we stand in the name of Jesus, we have the victory![14]

This spiritualization of a political struggle reassures believers that they *will* experience victory, if not in this world then in the next. It also justifies the current political inaction of these leaders; since they feel they have no control over when and how sovereignty can happen, it is better, in their view, to focus their energies on spiritual health and the healing of the community – necessary conditions for sovereignty. While Lloyd's hopes for sovereignty in his own lifetime have been diminished by the signing of the interim agreement, he feels reassured of victory for his people, and for himself, through spiritual action he sees as a-political.

During the dispute, the action taken – on the waters and in the community – was political. In hindsight, some people characterize that political action with religious language and emphasize the importance of the support of religion and spirituality. Lloyd and Leo are among the people who have said to me, "If only we'd held out for a little longer, the Canadian government would not have been able to stand in the face of what they did to us." Instead, the agreement has become another example of the injustice of the Canadian government (and the corruption of the Indian Act governments) carrying the day. For now, they feel exhausted with politics; they, along with many others in their community, risked everything for sovereignty, including their lives and their communities' safety, and now they see little change. They look for other ways to make a difference. Leo and Lloyd, and their families, now believe that the best avenue for change is not political but spiritual; prayer, anointing, healing, revival. They have transformed the issue of sovereignty into a spiritual one and believe that following a spiritual path is the way to political freedom.

From my perspective, the spiritual lives to which these people have turned remain political. They look for transformation in government – through prayer and through attempts to convert others to their views. They continue their connections to activist and advocacy groups. Their charismatic practice is a political position, one that might be seen as a "last stand" against intransigent Canadian and Indian Act governments. As Lloyd articulated earlier, his practice as a "follower of Christ" reinforces his identity as a Mi'kmaw man and leader. Within the group as a whole, during the year I attended meetings, conversation and prayer about the well-being of the larger reserve community were common, as was prayer for the transformation of political structures within the reserve and for the elected chief and council specifically.

After the dispute subsided, people continued to speak proudly of their actions in the dispute and link these actions to their religious lives. In the winter of 2005, I heard Leo give his "testimony" at the Pentecostal church in Tabusintac, where he was then a member. Testifying is an oral practice within Pentecostalism, the telling of the story of one's life and God's work in it. In his story, Leo spoke of the dispute, saying that in working for the community on the waters he was doing the Lord's work. This is particularly significant because his audience, the congregation, was entirely non-native, many from fishing families (both

English and Acadian) whose members would have acted against the
First Nation during the dispute. In his testimony, Leo characterized his
sovereigntist political activism on the waters of Miramichi Bay as work
for God.[15] He described how, while he was on the waters supported by
the prayers of his community, Satan was always there tempting him. In
Leo's characterization of that time, the temptations of Satan were anger,
rage, and violence; the work of the Lord was the work of enacting his
people's sovereignty and justice. His wife Audrey spoke often of how
the anger of the dispute was tamed and channelled by prayer. In conversation, Leo talked about the anger he felt during the dispute, and the
tension he felt between being sustained by anger and being sustained
by faith. Audrey explained that although Leo had a lot of anger at the
injustices they were experiencing,

> he was at the stage where he had faith that God was gonna protect him [on
> the waters and in the conflict]. Because he was praying at that time, he was
> getting more filled and filled each day as he went out there too. It wasn't
> anger any more after that; he was gettin' filled praying and acting for God
> and everybody in this reserve prayed. Everybody. Even myself, day and
> night, praying all day long. And, everybody was praying and that was
> what kept him safe.

Audrey doesn't distinguish the religious from the political here and instead emphasizes the importance of religion in her political life. In this
sense, charismatic "follower of Christ" positions are acts of resistance,
which allow sovereigntists to resist the Canadian government and, at
the same time, develop strong relationships and alliances with Canadian
charismatic people, with whom they worship and study.[16]

Rights and Nationalism in English Burnt Church

The argument that treaty rights were the reason for all of the upheaval
of the dispute did not appease English residents, or convince them that
what they were experiencing was justified. The catch of native fishers
was being sold openly on the wharf to local buyers.[17] Under the terms
of *Marshall*, to those who recognized treaty rights, this was how native
fishers could earn a "moderate livelihood," as they were allowed under
the court's decision. According to the DFO, this out-of-season selling
was illegal poaching. For many in the English community, this financial
exchange was simply open confirmation that the concerns of the

protesters were not about rights, but about money. One English fisher argued that "the ones that weren't fishing, but were fighting for it, I would say to them it was rights. The ones that were fishing, [it] was money. It had nothing to do with rights" (Mark). Most were not convinced by the argument that natives had been historically shut out of access to the fishery by the federal government and its policies. They recalled that many natives were lobster fishers before the 1950s, until the fishery went through a downturn, and they sold their licences and left the fishery: "That's why there was nobody fishing on the reserve. It had nothing to do with their right ... Nobody ever, ever, told them that they didn't have the right to fish, they had the same right as anyone else to fish – as long as they did it during the fishing season, the proper season" (Mary). These people were not convinced by the argument that native rights needed to be enacted to remedy the historic discrimination which had prevented natives from accessing resources in the past. They believe it was not discrimination that prevented natives from fishing, it was their own individual actions: leaving the fishery in hard times. In this view, treaty rights are not seen as a legitimate challenge to the authority of Canada's federal regulation.

Not everyone in the English community has a conservative view of native rights. There are some who acknowledge the historic tensions between the two communities and the separation between them, and who believe that their native neighbours have faced discrimination. These people wish for the possibility of some relationship between the two communities, some way to work together. They believe that increased employment on the reserve would be good for the people there, and for their relationships with the English community. But in the winter of 2004–5, when I was in Burnt Church, not many of these people felt hopeful that such a turn in the relationship between the two communities would be possible. Paul argued that the threat so many in the English community felt from the First Nation, and their resulting need to protect their own community, would prevent any real change. Perhaps, he said, something would change with another generation, "but overall ... I really don't see it in the near future." These people are willing to think seriously about native rights, but they also want the concerns of their own community to be taken seriously in the process. They have a moderate view – willing to recognize the experiences of their native neighbours, but wanting at the same time to ensure that their own community's needs are attended to.

Over time, the threats and protests of the dispute strongly affected the daily lives of the residents of the English community, as well as their views not only of their native neighbours but of their government, and eventually of themselves. This is illustrated by the mixed effects of the presence of the RCMP in the village. The RCMP's delayed response when they were called to a violent altercation between natives and non-natives in the English community gave many the sense that the RCMP were not really there to help them. This was reinforced by what some residents were told by officers assigned to their community during the dispute. On the one hand, the RCMP and DFO said they were there to protect non-native residents and the fishery; but on the other, they said that no protection was possible because they were not allowed to en-force the law with the native protestors past a very basic point. People began to feel frustrated by what they saw as a double standard: the laws and fisheries regulations of Canada were being enforced upon them but not upon their native neighbours. "If they're allowed to fish illegally, everybody should be allowed to fish illegally. They can't have one law for me and one law for them and one law for the next person ... If you're going to keep peace in the communities where you're so adjacent to each other, you've got to have one law for all" (Mary). The idea of a double standard illustrates that in the English village rights are seen as residing with individuals, within a larger world view that individualizes justice and responsibilities. People argue that Canada is on a slippery slope. This double standard, once begun in the fishery, will creep into other resource sectors (forestry, mining) and become an uncontrollable problem of greed. From the perspective of the English residents of Burnt Church, the government was not enforcing the laws of the land, since the protests and fishing continued. Canadian govern-ment agencies and elected representatives were no longer necessarily allies for local residents.[18] The shock of confronting their neighbours' deep dissatisfaction with the status quo was only compounded by the realization that their elected governments were not prepared to deal effectively with the situation. The dispute destabilized not only the lo-cal identity of these people, but their national identity as well.

For many in Burnt Church village, the problems of the dispute were problems created by the federal government. Some felt that the political and policymaking apparatuses of the government were responding to pressures from elsewhere in the country, rather than to the experiences and concerns of local voters. The decisions of the DFO, for example, have been shifted to Ottawa and out of the hands of local offices and officers. Some saw the local MP as responsive and concerned, but though he was a member of the ruling Liberal party, his was not a significant national voice during the dispute. Officials based in Ottawa were thought to be much more responsive to public opinion in Ontario, Quebec, and the West, where there are many seats and voters, rather than in northern New Brunswick, where the impact of their decisions was actually felt. "There's not too many voters here. What's in New Brunswick, six seats or something? So it doesn't really matter to them"[19] (Matthew).[20] This sense of the disconnect between local, sympathetic government employees and disinterested Ottawa bureaucrats was compounded by the duplicity that local commercial fishers and others saw in communications from the DFO and the RCMP. Commercial fishers would get information from local officials about catch rates in the native fishery, and then hear much lower numbers being released to the public. "What DFO was saying, to the public, is not what was in their memos. I have copies of a lot of it ... Most of it's blacked out, but you know what the feeling is. It was nowhere near what they were telling the public. Nowhere near at all ... It was two different agencies" (Mark).

This contradiction in the information that residents were getting from the government only served to reinforce the perception that the two different communities of Burnt Church were getting different treatment; and to the English community, it seemed that they were getting the short end of the stick: "The whole dispute is one race of people getting everything for nothing, and another race of people having to work and pay taxes, and hardly making a living at it" (Jake).[21] The people in the English community who took action to protest during the dispute did so because they believed that the Canadian government should be upholding one law for all people and that there was no basis for different treatment of Aboriginal people. They blamed the government for creating a situation of dependency on reserves by giving handouts to native people. The dispute was a profound threat to the continued prosperity of the English community, and it raised the spectre of their displacement from their waters and perhaps even their lands.[22] In this context, the defensive posture maintained by so many English residents is not surprising.

The perception that the government created the problems in Burnt Church has led to a renegotiation of the English community's relationship to government and its sense of national pride and identity as a Canadian community. People have an increased distrust of government services – especially policing – and of their reliability. Local residents believe that it has fallen to them to reinforce the Canadian identity of their place, where Canadian laws, values, and history continue to matter. For example, among some in the English village, there is a perception that the justice system is not fair or responsive to their needs and concerns.

> When are we going to get a fair justice system? [*Laugh*] Don't know when that would be … The justice system, I always thought before it was fair to everybody, but not anymore. It's crooked. Think if they got a new justice system out, or [one that could] be equal, like, with everybody, then it might change. But before that, I don't think you can get anyone around here to call the cops to come protect them, they protect themselves, I think. (Luke)

Many people hold this view: that there is no point calling the police or 911 if there is a problem, that you will get a more safe and satisfactory response by taking care of matters yourself. In one home I visited, a folksy hand-painted wooden plaque hung on the wall, decorated with a two-inch wooden machine gun. It read "Miramichi N.B.   we don't call 911." This dissociation from the government and its police is a specific breaking point, arising out of a broader sense of the mismanagement of the dispute by the government. This break seems particularly acute in a community with a high number of retired service people among its residents, people who have been engaged in Canadian judicial and enforcement structures throughout their professional lives. Linked with this anger and sense of injustice is also a sense of lamentation for things as they should have been. In an unpublished poem reflecting on the events of the dispute, local resident Marg Adamson writes:

> We are not important to the Government,
> And we are left here in the lurch.
> We are the forgotten people who are being sacrificed,
> The people of the other Burnt Church.

This sense of loss, and lack of support from the government, has not sparked an independence or sovereigntist movement, as has happened in other Canadian communities. Instead, English residents are trying to reclaim their community as Canadian. This is, perhaps, an important way to understand the initial protest of English residents at the onset of the native fishery in 1999, which resulted in so much violence. In the absence of significant government action, it became important for people to claim their place as theirs, as Canadian space, and their protest with Canadian flags and placards was the way they chose to do this. In the absence of a positive Canadian presence in Burnt Church,[23] residents continued to reinforce Canadian identity through public rituals and events.

During my year in Burnt Church, two major public events in the English community marked the significant contributions of residents to Canada – and marked the place as Canadian. The first was a major Remembrance Day ceremony, which included the installation of a cenotaph commemorating the contributions of veterans to the world and Korean wars and to Canadian peacekeeping efforts. The cenotaph itself lists the names of over 125 veterans from Burnt Church, the Burnt Church First Nation, and the neighbouring English village of New Jersey, and includes etched photos of five men killed in action in the Second World War. The installation of the cenotaph at the main crossroads of the English village on Remembrance Day included a parade of veterans from all three communities, and was attended by residents of all three communities. The cenotaph serves as a constant public reminder of local people's contribution to Canadian war and peacekeeping efforts.[24]

Crossing community boundaries to include native veterans in this memorial was a significant challenge for the locals of the English community who organized the cenotaph, and was made possible through the diligence of one interested non-native person. The monument sits at the heart of the English community, marking it as Canadian, and including people from both Burnt Churches within this demarcation. For English villagers, the installation of the cenotaph including native and non-native people was an important step in nationalizing all of Burnt Church, post-dispute, as a Canadian place.

The community celebration of Canada Day is also an important event in nationalizing the English village of Burnt Church as a Canadian place. In 2005, plans for Canada Day got started long before July 1st. In the late winter, the community got together and put on a variety show that packed the Women's Institute Hall, a successful fundraiser for Canada Day events. This was followed in the spring by a giant community garage sale, also organized as a fundraiser by the Canada Day committee. In the days leading up to July 1st, every home in the village

Figure 6. Burnt Church Cenotaph. The names of First, Second, and Korean Wars and Canadian Forces Peacekeeping Veterans are listed on this side of the cenotaph; "In Flanders Fields" is inscribed on the reverse. (Photo: Sarah King)

was festooned with Canadian flags, Maple Leafs, and red and white banners in anticipation of the celebrations. Canada Day itself was kicked off with a parade, followed by a picnic and games, with fireworks and a bonfire along the shore in the evening. The parade was a major event, passing all the important public spaces in Burnt Church. It started at St David's Church, turned at the old school and cenotaph, and then proceeded past the golf course to the wharf, where it turned around and headed back to the Women's Institute Hall. Almost every person from the English community was in the parade: children riding bikes, men driving fancy cars, the square dance club, banner carriers, the seniors' club, and others on homemade floats. In fact, participation in the parade is so important that few people were left to watch it. The parade was a community proclamation: This is a Canadian place! This is our place!

The creation of the physical structure of the cenotaph further cements the relationship of the settlers to their place, post-dispute. Casey has argued that "the very intertwining of culture and nature as it arises in oriented constructions specifies a fundamental aspect of place itself" (1993, 36). In this case, the construction of the cenotaph is intended to cement the relationship of the settlers to this place by invoking the names and images of those who risked or lost their lives in wars. The cenotaph reads, in part, "May we never forget their sacrifice / They served in many lands and returned peace to many nations." The invocation of this sacrifice, on a monument constructed by the community and installed at the centre of the village, is intended to sacralize the contribution of these men and women to their nation, and to this specific place. The cenotaph itself reclaims and sacralizes the relationship of the English residents and their ancestors to their place, Burnt Church, and defines the "sacrifice" of natives and non-natives alike as Canadian.

Basso argues that sense of place in "its social and moral force may reach sacramental proportions, especially when fused with prominent elements of personal and ethnic identity" (1996, 148), such as those expressed through the memorializing of sacrifice in war or through local cultural festivals such as the Canada Day celebrations. He suggests that communal relationship to place can be expressed and created through "recurrent forms of religious and political ritual" (109). Both the installation of the cenotaph at Remembrance Day and the Canada Day Parade were important political and religious rituals in the community of Burnt Church that endeavoured to re-establish a contested place as a

Canadian one. The military uniforms and protocols at Remembrance Day and the Canada Day regalia on people and houses function as important signifiers of identity; the parade marks out the relationship between people and place. The power of these rituals is, as Basso suggests, as sacraments of place; whether overtly, as in the installation of the cenotaph, or implicitly, as in the Canada Day celebrations.

American sociologist of religion Robert N. Bellah introduced the notion of a "civil religion" in 1967. By "civil religion" Bellah does not mean religion in general, nor some form of "national self-worship but [as] the subordination of the nation to ethical principles that transcend it in terms of which it should be judged. I am convinced that every nation and every people come to some form of religious self-understanding whether the critics like it or not" (1991, 168). Bellah argues that it is through "civil religion" that people attempt national self-understanding. In this sense, the public expressions of the English residents of Burnt Church at the cenotaph and on Canada Day – attempts to address the disjunctures in their self-understanding as Canadians resulting from the dispute – could be characterized as engagements with civil religion. People in the English Burnt Church are working to resolve the threat of their displacement and the lack of support from their government by enacting and ritualizing their sense of this place as a Canadian one, memorializing their historic commitments to (and sacrifices for) the nation, and nationalizing local places.

Quoting Kockelmans, Stefanovic suggests that a phenomenological value theory begins with "reflections on the foundations of morality ... at a level where the distinction between ontology, anthropology and ethics is not yet relevant" (2000, 124). Such reflections assist us in understanding the pre-theoretical commitments and experiences upon which moral and ethical arguments are built, in a way that resists universalizing and relativizing tendencies (122). The dispute in Burnt Church highlights the problem of contested places. In a place inhabited by two separate communities with different ties, histories, and relationships to place, this contestation is negotiated, at least in part, by a turn to questions of rights, sovereignty, and nationalism. In this context I ask, "What do the various engagements with rights, sovereignty, and nationalism illuminate about relationships to place in Burnt Church, and about the values arising within this place?"

The importance to each community of re-establishing their authority over this place is apparent across these engagements. Authority, for

both of the Burnt Churches, is not simply authority over land or ownership of territory; authority is, perhaps, the way of guaranteeing the pri-
macy of a particular way of being in a particular place. Nationalist and
sovereigntist positions and arguments from and for rights are claims
for the authority of particular ways of being and thinking in place. If we
understand concerns for authority in Burnt Church/Esgenoôpetitj as
reflected in these arguments for rights, sovereignty, and nationalism,
we begin to see that the meaning of authority itself is understood differently by the two communities in this contest.

Within the English community, instantiating the Canadian identity of
Burnt Church is a way to reclaim and re-characterize people's relationship to their place, post-dispute. The re-establishment of the authority
of "Canadian" values, identity, and sense of order is a critical concern.
Such values seem to include the importance of equal rights for every
individual, and the rule of one law for all; these values were expressed
earlier in this chapter in criticisms of the departure of the Mi'kmaq from
the commercial fishery, and in the enforcement practices of government
agencies during the dispute. The value of labour in establishing ownership and legitimacy, and the importance of hard work as the guarantor
of reasonable survival, if not success, is expressed as a concern for in-
equality. The "threat" to the English community that arose during the
dispute, which appears as a threat of violence, is more importantly the
threat of loss of authority over their place, and of dispossession from
the fruits of their labours. The ongoing arguments for "Canadian" values and order is the framework for articulating concern for equal rights
for individuals, one law for all, and the role of hard work in earning
livelihoods and establishing authority.

Within the First Nation, reclaiming rights and acting for sovereignty
is a way to restore what has been lost, where possible, and a way to agitate for just and rightful relationships with the Canadian government
in the future. The concern for rights is an expression of the value placed
on right relationships, within the Mi'kmaw community, between the
Mi'kmaq and the Canadian government, and among the Mi'kmaq,
their ancestors, and the other living beings in this place. Arguments for
the importance of the re-establishment of traditional Mi'kmaw politics,
culture, and authority in this place draw upon these traditional ways of
life as the foundation of the healing and re-establishment of proper relationships in this place, at all levels. As we have already seen, some in
Esgenoôpetitj suggest that the attempts that were made to re-establish
this authority, and the relationships that it fosters, prompted healing in

the community. Right relationships are interconnected with other community values and priorities, including justice and healing both for people and their place.

These two competing sets of values, senses of place, and claims of authority are expressed in rights talk, and also as concerns for conservation. For both communities, the rights-based or nationalist positions of the others can open possibilities for relationship, as in the inclusion of all groups of veterans in the cenotaph. More often these positions are seen as threatening, and close off opportunities for further rapprochement. As chapter 5 elaborates, the global discourse of conservation – which seems (to non-locals) to demonstrate concerns and experiences common to the communities of Burnt Church/Esgenoôpetitj – often served in the dispute to reinforce division and difference between the two. The question remains, as people in the Burnt Churches negotiate place and identity post-dispute, whether the turn to healing, Mi'kmaw tradition, and charismatic Christianity, or the turn to Canadian nationalism, can provide a foundation for shared concern, or whether they will become static divisions, reinforcing the opposition of the two Burnt Churches.

5 Conservation Talk: Negotiating Power and Place

In 2002, after many months of consultation, the report of the Miramichi Bay Community Relations Panel's investigation of the conflict in Burnt Church was released. This report documented many of the concerns that local natives and settlers had had about the dispute. It was this report that prompted the chief official of the Department of Fisheries and Oceans (DFO) in the region to comment, "Perhaps it never really was about fish" (CBC 2002). As I have shown, the issues at the heart of the dispute are issues of place and displacement, indigenous rights and sovereignty, and authority and belonging. Yet during much of the Canadian government's dealings with communities involved in the dispute – in its public positions and in the parameters it set for mediators such as the Miramichi Bay Community Relations Panel – it persisted in defining the dispute as being "about fish." The government positioned access to and conservation of the lobster fishery as the key issue to be negotiated and discussed, and disallowed other topics from the conversation.[1] For people in both Burnt Church communities, this meant that some of their most important concerns would remain unaddressed unless presented as a dimension of conservation.

In previous chapters, I demonstrated the importance of place in the dispute, and of religion, values, and identity as dimensions of place, though these concerns were often implicit or unacknowledged in the public discourse. The discourse of conservation illustrates local attempts to explicitly articulate their concerns for place in nationally acceptable terms and frameworks. This exploration of the meanings of conservation is an attempt to get at the implicit values being contested in the dispute; it also represents an attempt to look critically at the shared concerns and experiences of the English and Mi'kmaw communities of Burnt Church. Both communities engaged in conservation talk to express their concerns for livelihood, and their critiques of the federal government. In chapter 6 I will look more closely at the calculative and managerial sense of place driving the work of the Canadian government within the dispute. In this chapter, the focus remains on the discourse of conservation within Burnt Church and Burnt Church/Esgenoôpetitj.

During and after the dispute, the conservation discourse reflected not only the concerns of the Burnt Church First Nation and village – such as access to livelihood – and their critiques of the federal government, but also attempts by both groups to win allies to their respective positions. The role of this discourse in the communities must be understood as a dimension of both the historic processes of colonization and the ongoing marginalization of indigenous and rural communities through globalization. Though it can tend to silence local values and priorities, the global conservation discourse also became a critical tool and framework in the attempts of the Esgenoôpetitj First Nation and the English village of Burnt Church to get at least some of their concerns on the table. Interestingly, although the divisions remain between the two Burnt Churches, the discourse of conservation also reveals values shared by both communities.

Colonization, Globalization, and the Discourse of Conservation

In the philosophical literature on place, the archetypal stories of violence done to place are almost always stories of the displacement of indigenous people through colonization. Casey chronicles the devastating impact of the twentieth-century displacement of the Dineh (Navajo) from their traditional lands and territories (1993, 34–9). Malpas begins *Place and Experience* with examples of the significance of place to Aboriginal Australian and Maori peoples (1999, 2–4). Stefanovic argues that perhaps no societies have better understood the power of place than indigenous or Aboriginal societies (2000, 113). Basso's exploration of place within the White Mountain Apache has become *the* ethnography of place for many authors, including myself. Basso is not the first to argue that the significance of place is often taken for granted and only becomes apparent when we are deprived of these attachments (1996, xiii), as indigenous people have been through colonization. Authors such as Casey, Malpas, and Stefanovic draw on these insights from indigenous cultures in their exploration of place and displacement and their significance for contemporary Western society (and philosophy). Casey points out that "the sufferings of contemporary Americans ... uncannily resemble ... those of displaced Native Americans, whom European Americans displaced in the first place" (1993, 38). This resemblance is a critical question in the discussion of place and displacement. It is not that settlers are *like* indigenous peoples, or that there is some sort of convergence over time between the two cultural groups, erasing difference. Rather, the same *processes* that subjected so many Aboriginal people to displacement are now also displacing some within settler society. What was visited by settlers on indigenous societies through colonization is now reproduced in a more moderate form within settler societies with the expansion of globalizing, colonizing modern powers.

Colonization is a process done to places, and not simply to people. In Canada, one of the primary projects of the colonial governments of Britain and the early Dominion (as Canada was known after 1867) was the signing of treaties with native people for rights to land and landscape. Over time, the colonial government in Canada removed indigenous people from their lands and disrupted their traditional forms of livelihood in, and relationship to, place.[2] Casey has argued that place is a cultural complex arising from the epicentres of body and landscape, that place consists of embodiment, landscape, and culture (1993, 29). It was upon these three dimensions of indigenous life that colonialism was focused. Colonialism acted upon indigenous bodies through disease, through laws, and through the physical removal of native children to residential schools for "re-education." Colonialism effected the removal of indigenous peoples from their traditional lands and landscapes; the outlawing and regulating of traditional practices; and the separating of indigenous cultures from the places in which they had meaning. Appropriating indigenous places was *the* primary purpose of colonization.

Historically, this process of appropriating place, and re-inscribing it as a colonial landscape can be seen in colonized places around the world. In Egypt, Mitchell has described the colonial project as one that "inscribes in the social world a new conception of space, new forms of personhood, and a new means of manufacturing the experience of the real" (1991, ix). Trouillot explores the making of Haitian history as a process of power and colonization that silences people and remakes their places in the terms of the colonizers (1995). In India, Guha traces the roots of contemporary peasant resistance movements to the colonial period, arguing that Eurocentrism shapes contemporary globalized forestry in forms continuous with its power in colonial times (1989b). He goes on to suggest that Western conservation movements, especially those inspired by deep ecology or concerned only with saving large fauna (such as tigers) but not people, are equally problematic in that they represent another effort to impose Western values on a globalized south (1989a). In a very important sense, the discourse of conservation, though positioned as a resistance to the powers of globalization, is in many ways itself a dimension of the globalizing discourse. The processes of globalization and development are modern incarnations of colonialism, displacing indigenous and other marginal peoples in favour of capital and conservation.

In North America, the connections between colonial and global powers are even more clear, because our globalizing nation states are not post-colonial; settlers have not left, as in India or other places in the global south, but remain in power and in residence. The significance of the ongoing presence of colonizing and globalizing powers and governments has been explored by many indigenous authors in the Americas, including Thomas King (2003), Taiaiake Alfred (2005), and Vine Deloria (1994). Historic colonial efforts to appropriate place continue in the present through the development of a globalized economy (led by colonial powers such as the United States, Canada, and Britain) and a globalized conservation discourse (led by conservation groups headquartered in these and other countries). For indigenous peoples, the dual oppressions of globalism and colonialism combine in the conservation discourse to recreate and romanticize indigeneity. In those few wild places that have been "granted" to indigenous peoples, natives are supposed to live as the epitome of deep ecological and conservationist principles, a living museum of subsistence behaviour, while the rest of us go on about our business on the rest of the land (Francis 1992, Kretch 1999). In this view, indigenous people do not desire or need to participate in the larger economy. This romantic view of indigeneity can be used to advantage by indigenous peoples seeking to mobilize the sympathies of non-native allies. This can be seen in the adoption of traditional dress by Cape Breton Mi'kmaq for political appearances against Swedish forestry corporations in Sweden, as Hornborg has discussed (1998), or through the presentation of indigenous Amazonian identity in popular forms to mobilize allies, as Conklin has explored in great depth (1997, 2002, Conklin and Graham 1995). At the same time, as Dombrowski has shown in Alaska, the conservationist discourse serves globalizing corporate and national interests by reifying and limiting what it means to be indigenous (2002).

In the Canadian context, the power of conservation as a legal and political tool in the negotiation of relationships with Aboriginal people has been reinforced by the Supreme Court in *Sparrow* and *Marshall*. In *Sparrow*, the court recognized the Aboriginal rights of BC Aboriginals to fish, while outlining the conditions under which these rights can be regulated and limited. The court held that conservation is the only thing more important than Aboriginal rights in management decisions (see Notzke 1994, 56). In *Marshall II*, the court affirmed that for treaty rights, the "paramount regulatory objective is conservation" (R v. Marshall 1999b, 3). In this case, the court went on to say that responsibility for conservation "is placed squarely on the minister responsible, and not on the aboriginal or non-aboriginal users of the resource" (3–4). The legal positioning of conservation as the government's trump card reinforces the social and political power of the conservation discourse within Aboriginal communities (King 2011). In Esgenoôpetitj, where community concerns about rights and sovereignty found little purchase in the public debate, the discourse of conservation became the framework within which political positions were articulated and contested, allies sought and opponents challenged. While the community recognized the importance of conservation as stated by the court, they did not accept that this was solely the responsibility of the minister, as I will explore below.

The modern effort to colonize and globalize places has led to the displacement of non-indigenous people – settlers – on two fronts. As many philosophers of place have pointed out, people in modern Western societies experience a sense of placelessness because "all modern places are essentially the same: in the uniform, homogenous space of a Euclidean-Newtonian grid, all places are essentially interchangeable" (Lassiter in Casey 1993, 38). This is the McDonald's-ization or Walmartization of our communities, when the goal of development is to reproduce precisely the same location and experience in every landscape. On a social and cultural level, we in the West are encouraged to seek this uniformity of place. More significantly for our study of the Burnt Churches, we have begun to marginalize communities *within* our society when their relationships to place do not fit within the dominant homogenizing paradigm. Rural communities are fundamental misfits because they often have a relationship to place that resists homogenizing. As Vandergeest and DuPuis have argued in their introduction to

Creating the Countryside, rurality is often constructed as if it is in opposition to urban life: rural communities are natural, peripheral, and in the past as opposed to the cultural urban centres of the present (1996, 3). The meaning of rurality is constructed largely by dominant groups outside of rural communities, while rural people themselves are marginalized because their self-understandings do not fit within these pictures of rural life. Political and economic powers, and those who challenge them such as environmental groups and NGOs, are all situated in cities, and while their accounts may oppose one another, neither group usually attends to the perspectives of rural people themselves (Vandergeest and DuPuis 1996, 6–7).

The conservation discourse coming from urban power centres often continues to link rural communities with nature – as communities of the past and separate from society – a conservationist "vision that saves and purifies nature by eliminating the social, including local histories of human activities" (Vandergeest and DuPuis 1996, 14). In this view, natural spaces are supposed to be "free" of people, other than the few subsistence-dwelling Aboriginals discussed above. The complex relationship between rural people and place is overlooked in favour of the ideologies of conservation: ideologies that divide the human from nature and the urban from the rural. Rural communities that were once the agents of colonization, transforming the landscape and recreating indigenous places as their own, are now finding their lives and livelihoods under threat from the forces and governments they were a part of. In a U.S. example of this phenomenon, Devon Peña explores the importance of long-standing community water networks (*acequia*) in the villages of land-grant Hispanic communities in Colorado and New Mexico. He describes how the enclosure of the commons and the resulting dramatic timber harvest, the "degradation of [their] homeland by the forces of modernity and maldevelopment" (2002, 61) has changed their place so profoundly that the people feel it as *susto*, or loss of soul (66). In Burnt Church, the residents of the English village find themselves facing forces of industry and government that use the conservation discourse to gloss over the needs and concerns of their community, while in conflict with indigenous neighbours who are seen much more favourably by environmentalist and NGO actors. They respond to this challenge by finding ways to characterize their concerns for livelihood, as well as their critiques of the government and the Mi'kmaq, within the discourse of conservation so that their needs might be received in a more favourable light.

Casey has suggested that the solution to our modern problem of place lies not in nostalgia or exoticism, but perhaps "in a belated postmodern reconnection with a genuinely premodern sense of place, a sense such as the Navajo once had and may lose altogether unless something is done to restore them to their land" (1993, 39). The suggestion that indigenous communities are under threat and at risk of extinction is implicit in Casey's argument. As Peña has acknowledged, in present circumstances it seems easy to "confirm a prognosis of a disappearing culture" (2002, 71), much as the early colonists of Canada did. But the displaced cultures and communities of Aboriginal people have not disappeared; they are engaged in a vibrant regeneration and resurgence across the country. Displacement in modern times is a political problem, which cannot be solved simply with a reconnection to a premodern spirit of place. Certainly, in individual places, people and communities engage in the re-inhabitation of place in ways that are religious or involve public ritual, as discussed in the preceding chapter. Yet these same communities, like the Burnt Churches, are simultaneously developing positions of political resistance that are clearly linked to their place-based and religious identities. Indigenous senses of place are qualitatively different from rural senses of place, as the experiences of colonization within indigenous communities are different from the experiences of globalization in rural communities. What communities such as those in Burnt Church share is that, in the present, they are both marginalized and displaced by similar processes of power. While the Mi'kmaw residents of Esgenoôpetitj have a long and complex experience as the objects of colonization, members of the English community whose ancestors were the agents of colonialism on these lands now find themselves objects of its new, urbanizing, globalizing form.

In this context, the discourse of conservation becomes an important political tool for the residents of both Burnt Churches. During the dispute, the members of each community strove to articulate their concerns and justify their positions in terms that would be easily and sympathetically understood by outsiders and familiar in global discourse. The conservation of the lobster fishery became a key framework for these arguments. In the preceding analysis, I have suggested that the greatest threats to the sustainability and implacement of these communities lie in colonialism and globalization. Throughout and after the dispute, the residents of the two communities continue to see each other, and the government, as significant threats. Their conservation discourse draws out both of these targets, using a globalized language of conservation to represent local interests. Conservation is a framework through which they can make their concerns and values more convincing and acceptable in the public context, and reclaim and re-instantiate their power in their particular place. For some, conservation is necessary to preserve (or to create) their livelihood. For some, conservation is grounded in a critique of government. For some, conservation calls for, or is code for, social controls exercised upon their communities; articulating a conservationist position is a way to seek allies in resistance.

Livelihood

Conservation is not primarily important in English Burnt Church as a value in itself. Nor do locals have a profound connection or identification with the fish stocks, as suggested by deep ecology. Such ideas were never mentioned in any conversations during my year in Burnt Church. Conservation is employed as a way for people to protect their own livelihoods. Fishers are trying to protect their own access to the resource so that they can feed their families, pay their mortgages, and send their children to university. The importance of livelihood cannot be overemphasized in the small resource-based economy of the Miramichi. Burnt Church offers few possibilities for employment. Many work fishing lobster – supplemented by herring, mackerel, and oysters – on crews for small inshore boats largely run by extended families. Some supplement their family income with work in the forests, especially in the fall, when "cutting tips" for Christmas wreaths and garlands, though piecework, can add significantly to a family income. Until the early 1990s, people could find employment at CFB Chatham and a local mine, but these have both closed. Some in the community were in the Canadian Forces and moved home to Burnt Church upon retirement. A few work at the local paper mill, and these millworkers are seen to be the most fortunate, as they have unionized jobs with pension and benefits,[3] as do the local teachers and employees of the regional maximum-security penitentiary. Finally, the small Burnt Church Credit Union, open three days a week, provides part-time employment to one person. Almost every family in the English village of Burnt Church relies on some form of resource-based employment to meet its needs. In this economic context, inshore fisheries like lobster are critical for the survival of the community as a whole, representing the only sustainable employment available right in the community. Much like the cod to the Newfoundland outport, the lobster is at the heart of village life in Burnt Church: it is the foundation of the livelihoods that permit this community to continue, and sometimes to thrive.

During the dispute, the direct conflict between commercial fishers and (unlicensed) native fishers often occurred on the water: non-native fishers would go out in their boats and cut the traps of native fishers. English fishers and locals insisted that they did not cut traps, that the Acadian fishers from neighbouring communities, especially those across the bay, were responsible. At the same time, English residents shared the concerns that motivated the trap cutting: they were concerned about the effects of the native fishery on lobster stocks in the region, the possible collapse of the fishery, and the loss of their livelihoods. The spectre of stock collapse was powerful for the English, who see the native fishers as a new group of people accessing the limited fisheries resources without the normal limits of government regulation. Most of the English commercial fishers that I spoke with talked about the thousands of traps in the waters during the dispute and the millions of pounds of lobsters being caught "out of season." They perceived the native fishery as large and unregulated, operated by people who did not have strong skills or knowledge about fishing lobster. For some English fishers, such as Mark, this perceived lack of knowledge was the only thing that mitigated the threat of the native fishery: since the natives were not considered to be good fishermen, they were apparently less likely to have large catches and threaten the stock.

> If they [native fishers] hadda known where the good spots were, then it woulda been a lot worse. See, that's why the fishermen were so mad … Now last year [2004] there were no lobsters. Whether there will be this year or not, nobody knows. That's what you were trying to prevent. (Mark)

At the same time, the sheer number of traps alleged by the DFO to be in the water and the number of traps counted by English fishers when they went out in their boats were seen to be a real and serious threat to lobster stocks, one that has had an impact on the subsequent spring commercial fishery.

For English fishers, the other serious threat to lobster stocks was the DFO's ineffective enforcement of commercial rules and regulations. The view from the English village was that the DFO was not telling the whole story about how large the native fishery "really" was. The DFO's perceived downplaying of the impact of the native fishery on lobster stocks was seen by English fishers as self-serving in the short and long term. In the short term, the DFO was perceived as "not wanting to get involved" with the challenges inherent in regulating the native fishery. In the long term, some fishers were suspicious that the DFO was using the dispute to further its own agenda of reducing the number of licences and boats (i.e., people working) in the lobster fishery. Mark argues: "One thing is gonna be that the DFO is going to have exactly what they wanted, twelve years ago ... 30% less fishermen. The sad thing is it's gonna be all the young fellas [that go]." Others saw the actions of the DFO in the dispute as a part of the ongoing incompetence of the agency, which will, it is feared, lead to stock collapse in the inshore lobster fishery in the near future, much as it led to the earlier collapse of the cod fishery in Newfoundland.

Fishers and locals are almost uniformly pessimistic about the future of the fishery post-dispute. This pessimism is compounded, in the eyes of English fishers, by a third problem. In order to get natives into the fishery under the agreement-in-principle, the federal government acquired existing lobster licences to give to the First Nation. These licences were purchased from lobster fishers along the New Brunswick coast and transferred into the Burnt Church zone. The result was a higher concentration of fishing in the immediate area, even though the number of licences in the overall region did not change. According to some fishers, in the seasons leading up to the dispute they had begun to see a slight improvement in stocks as the result of their own conservation measures implemented in the 1990s (after *Sparrow* allowed a fall native food fishery). They believe that this success is lost because of the native fishery during the dispute and the locally increased commercial fishery resulting from the settlement of the dispute.[4] In all cases, it seems to the English fishers that their fishery is as vulnerable to collapse from government mismanagement as it is to collapse from overfishing.

In English Burnt Church, the conservation of the lobster stocks is important because of their use value to humans – but not only this. The continued existence of the lobster fishery is critical to the lives of people on the land and waters. Conservation of jobs in the fishery is important, not only to individual fishers, but to the life of the entire community in this place. The concerns that English people express for conservation are certainly concerns for their own continued livelihood, but they are also implicitly concerns for the extended community of which they are a part. Concern for livelihood is an intrinsic part of conservation in this place, for the sake of individuals and the community as a whole. When

English residents argue for conservation, they are arguing for the future of their community and a notion of conservation that takes humans and nature together, and resource-based livelihoods, seriously.

This interest in livelihood is not only an interest of the English fishers. The folks in the English community largely *have* livelihoods that they are trying to protect. The folks in the Burnt Church First Nation largely do *not* have livelihoods. In 2000, the on-reserve unemployment rate was at 85 per cent (Dharamsi 2000). The only stable local employment is with the band council and its programs. Prior to the dispute, there was one commercial fisher in the community, but the entry costs for licences and gear were prohibitive to others. A lobster licence, for example, was listed for sale at between $300,000 and $450,000 in August 2007, though the prices change over time and according to the zone in which the licence is held.[5] When there are large government buybacks of licences, such as the one to get licences to give to natives at the end of the dispute, the price goes up. Since credit is largely not available to residents of the reserve (they have little collateral because all property is held communally, according to the Indian Act), these costs are a barrier to independent entry into the fishery.

The action on the waters in the dispute was, for some, an attempt to find a way to earn a living to support their own families. Miigam'agan, a Mi'kmaw woman and community leader, emphasizes this when she speaks of her sister, one of the first to put traps in the water after *Marshall*:

> My sister called … and she said that she wanted to go out and fish. She is a single mom, with two children, and she had not had employment in a long time in our community. And she went to the band office and pleaded … if she could get her welfare cheque early, you know … What she did with her welfare cheque is that she invested and got – I think she got 20, 25 traps, I can't remember. And she didn't have a boat, but she had made contact with another boat owner in the community and asked them to take them out for her … She got these old wooden traps, she got bait, all the things that they told her she would need … She was really excited when she made her first catch, and she was selling her catch, to buy bait so she can continue to fish. Now she had enough to buy the bait, and help out with the fuel for the boat.

This is a poignant story, and I'm sure that Miigam'agan told it to me because of what it says about gender and about power in her

community; but it also illustrates something important about livelihood in this dispute. People in Esgenoôpetitj saw an opportunity after
Marshall to earn a living for their families on their own terms and remain in the community. In an isolated politicized northern reserve, that
is no small feat. In chapter 2, another community member described
how, as DFO raids and trap seizures increased, native fishers put "little
letters in the traps, little toys for the kids, stuff like that" to get the importance of livelihood in their community across within the fishery. In
practical terms, in Esgenoôpetitj, the right and opportunity to earn a
livelihood was a key motivator for people involved in the dispute and
an important dimension of the community's articulation of the need for
conservation.

People in each of the Burnt Churches view the attempts of residents
in the other community to earn a livelihood with jealousy and distrust.
Each community thinks that the other has it "easy." In the First Nation,
many people see the relative wealth and prosperity of their English
neighbours, and they suggest that it is built through access to resources
denied to the Mi'kmaq on land that was once theirs.[6] In the English
community, many people see the government's obligations to the First
Nation as "handouts" that are not appreciated or properly used by native residents. Though livelihood is a common problem in each community in different ways and for different reasons, there is little common
ground between the two communities to work together towards this
goal. I have argued earlier in this chapter that these two communities
hold similar marginal positions with respect to the larger problems
of globalization, but this commonality is something few residents
recognize.

Critiquing the Canadian Government

The dispute was (and is) not just about fish. The dispute is, in one sense,
about people's ability to earn a living for their families in their home
communities with a limited resource. In another sense, conservation is
a political discourse. In both the English village and the First Nation,
conservation talk is grounded in critique of the federal government and
its policies. In Esgenoôpetitj, especially for the sovereigntists, the DFO's
history of fisheries mismanagement demonstrates the importance of a
Mi'kmaq-managed fishery. In English Burnt Church, the critique of the
DFO focuses not only on mismanagement but on the disengagement
of the government from the settler communities it is "supposed" to

represent. In both communities, people are suspicious of how ideas of conservation are used to exercise control over their communities, and they are working to turn conservation into something that serves rather than subordinates them.

After the *Marshall* decision, members of the Burnt Church First Nation engaged in a community consultation process out of which arose a community management plan for the fishery, *Draft for EFN [Esgenoôpetitj First Nation] Fishery Act* (Ward and Augustine 2000a). The fisheries management plan was endorsed by the Conservation Council of New Brunswick, and those invested in it believe that it represented a much more conservation-friendly approach to fishing lobster (and other stocks) than the management plans of the DFO. During the later years of the dispute, the native fishery was carried out according to the management and conservation principles of this plan, which had been approved by the community as a whole. The plan itself is quite critical of the Canadian government, and of the DFO in particular:

> Directly due to DFO[']s economically focused management plans there are now over 500 species of fish in the Atlantic/Quebec region that are at risk.
>
> The focus of the fishery management by the DFO was not to protect and preserve the fisheries and it's supporting ecosystem [*sic*]. DFO[']s focus was to satisfy the non native fishing industry and ravish the fisheries for the sake of profit. This policy has been at the expense of the Mi'kmaq, Maliseet and Passamaquoddy fishery. The DFO have historically forced the Mi'kmaq, Maliseet and Passamaquoddy people out of their own waters and denied them their inherent rights so the DFO could selfishly take over the fishery and make non native fishermen wealthy.[7] (Ward and Augustine 2000a, VII)

The failure of the DFO to successfully manage the fisheries (i.e., to ensure a thriving, stable resource available to all) is, in this view, not simply due to incompetence, though clearly the authors believe the DFO to be incompetent.[8] From the perspective of the Mi'kmaq, it is clear that the fisheries policies of the government were *intended* to disenfranchise native fishers for the sake of the non-native fishery.

The native fishery in Burnt Church is framed by the Mi'kmaq as a conservationist response to colonialism and to the incompetence of the Canadian government's management of fish stocks through the DFO. Through the plan, the people of Esgenoôpetitj claim a fishery as their traditional right within their traditional territories. The conservation plan of the community is also a clear political statement.

> Due to the consistent mismanagement by DFO, it's biased and racist policy making, it's overpolicing of Mi'kmaq fishermen, it's adversarial nature and relationship with the Mi'kmaq, it's paternalistic and condescending attitude towards First Nations people, the Mi'kmaq of EFN [Esgenoôpetitj First Nation] will be reasserting it's control over the fisheries in it's traditional territories.
>
>   The EFN will exercise its Inherent right to self determine it's own political, social and economic future and it's inherent right to self government which will include the ability as a self governing people to legislate policy. (Ward and Augustine 2000a, VII, punctuation as in original)

In *Sparrow*, the federal government was charged with developing management of the BC fisheries in cooperation with BC First Nations. This co-management, as it developed through joint stewardship agreements in the 1990s, had mixed success. The Gitksan Wet'suwet'en, who pursued fisheries management as a part of their larger goal of "ownership and jurisdiction" over their territories, found the process especially conflicted. In 1992, the Speaker of the Gitksan Hereditary Chiefs suggested that most joint-management agreements contain two clauses that make them untenable for First Nations people, the "God clause" and the "greed clause": "the Minister has the final say" and "everything has to make economic sense" (Ryan in Notzke 1994, 59). While the Gitksan experience of co-management is perhaps more conflicted than some, the troubled history of co-management in the BC fisheries may have informed sovereigntist politics in Esgenoôpetitj at the end of the 1990s.

For sovereigntists within the First Nation, the wealth that Canada is concerned with conserving is earned from stolen resources. As the traditional chief put it to me, "What makes you think it's your moneys, it's my resources that you're playing with?" In this view, Mi'kmaw management of the fishery is a necessary condition of effective conservation; effective conservation encompasses sovereignty and justice. The fishery management plan of the First Nation represents the position of the community as sovereigntist – and anti-colonial. Through this document, the community argues that the federal government and its agencies pursued policies intended to separate them from their lands and resources and that its protest (through the fishery) is a reasonable response, an attempt to reclaim what is rightfully theirs under the treaties. Meeting the concerns of livelihood and conservation in and of themselves is not enough; they need to happen on Mi'kmaw terms in a Mi'kmaw nation. The Mi'kmaq argue that while the federal government is not able or willing to be conservationist, they are.

The *Draft for EFN Fishery Act* opens with the language of rights and anti-colonial resistance and concludes with an emphasis on conservation. Access to the fishery will be granted on the basis of a "conservation priority system," in which the ceremonial, food, and social fishery take precedence over the activities of commercial fishers (Ward and Augustine 2000a, XXI). The plan goes on to suggest that the fishery will be guided by a developing "Mi'kmaq conservation philosophy," based on scientific data, traditional environmental knowledge from Mi'kmaw fishers, and traditional philosophy from elders and community members (XXIII). The overall intent is to restore the habitat degraded by DFO mismanagement. The act suggests that the dollars earmarked by the DFO to buy native fishers into the regulated commercial fishery (buying licences, boats, and constructing a wharf) should instead be used to retire existing commercial licences, reducing the overall size of the non-native fishery "for the sake of conservation" (XXV). The political nature of the conservation question is acknowledged within the plan itself, with the authors suggesting that the DFO and the government will use conservation to create conflict between native and non-native fishers.

> We are very concerned that the DFO will attempt to politicize the current conflict between EFN and DFO. Instead of immediately complying with the SCC Marshall decision and providing "access" for members of EFN, we believe the DFO will try to use our management plan as a means of creating a "conservation scare" amongst non-native fishermen. (XXV)

Conservation is a political framework, which members of the First Nation use to contest the government's position on their fishery. It is recognized here as a tool that may be used by the government to drive a wedge between local communities with interests in the fishery.[9] "Conservation" is also the grounds by which the First Nation seeks alliances with non-natives against the Canadian government, through an explicit invitation in the policy to those who "share the same conservationist principles," a point that will be explored further in the next section of this chapter (XXIV). The members of the Burnt Church/Esgenoôpetitj First Nation, through their management plan, are engaged in a political contest over who is the most capable of implementing sustainable management of the fisheries. Like any policy document, the EFN management policy as laid out in the Fishery Act is a *political* document. The act articulates issues of rights and conservation in an attempt to discredit the Canadian government's fisheries management and position the Mi'kmaw fishery as conservationist in principle and in practice.

English residents also criticize the federal government and its departments for more than simple fisheries mismanagement. People within the English Burnt Church often suggested to me that the government did not exercise sufficient control over the native fishery quickly enough. Underlying this critique is the sense that the government never did have the interests of the English village in mind. One fisher argued that the source of the problems with fishing, including declining stocks and conflicts on the waters, lay in the removal of decision-making power from the local DFO offices and officers to bureaucrats and scientists in Ottawa.

> Decisions about the fishery, in the last 10 years, have all switched to Ottawa ... Mostly because of the native issues ... It's not the fishery officers in the field, it's not their fault. They only do what they're told. They probably would have cleaned the mess up. But they weren't let do it ... It was all coming from Ottawa, so. And as long as they're up there they get to see it in the paper, or on the news, and they get reports. They only have to read what they want. (Mark)

Decision makers who work at a complete remove from places they are supposed to manage are at a serious disadvantage, because they do not have the opportunity to develop an understanding of the local context. Such challenges are normal, or commonplace, in the contemporary globalized context. Peña raises the problem of decision making at a distance in his discussion of acequia communities of the American South, where political and bureaucratic decisions are made from Washington. In that context, locals wonder how "policy can be fashioned by people in a place as far away as Washington DC, by people who have never been on the land" (Peña 2002, 65).

For the residents of English Burnt Church, this distance and disengagement on the government's part gives rise to a complicated relationship with Canadian government and laws. Unlike the Mi'kmaq, the English generally expect the Canadian government and its agencies to represent and engage their interests. The distance of the agencies of government from the people they are supposed to serve represents a powerful and intransigent problem for the people of Burnt Church, as they look not only for solutions to the dispute but also for ways to manage the fishery sustainably in the long term. When the government fails to understand local interests, as it has been perceived to do in the case of the DFO (since well before the dispute itself), some English residents continue to look for some way to repair or maintain the system so that it might serve their own needs.

Others go ahead with life on their own terms, believing that any possibility of successfully conserving the fishery lies in local hands and practices. This position is echoed among non-native fishers in the region, where, since the dispute, violence has again taken place on the waters (2005–7). The post-dispute conflict, between commercial fishers at different wharves, is over the interpretation of traditional (unofficial) fishing practices, specifically the location of lobster traps at different times of the season. Post-dispute, the importance of unregulated local fishing practices, and their perceived impact on livelihood and the resource, has heightened. Among fishers who were already independently inclined, the importance of local fisheries management traditions has increased with their declining trust in the government. Residents of non-native fishing villages do not believe the DFO serves or represents them. In this view, the people best able to conserve the resources of the fishery are non-native fishermen, who are seen to have the necessary combination of skill, experience, and interest in maintaining the fishery.

Beyond the political critiques of the government enumerated above, some within the First Nation offered a deeper analysis of the meanings of conservation for the Canadian government, a significant indictment of the government's motivations in the dispute, and an excavation of the hidden values at play in the government's discourse of conservation. From this perspective, the government and its agencies – specifically the DFO and RCMP – used the language of conservation to justify actions that primarily preserved their own power and standing. In the government's case, some suggest, their interest in maintaining social control trumps all other interests. Cindy, a Mi'kmaw woman from a fishing family, argued emphatically that the federal government used conservation talk to promote its own interests:

That's the biggest word that they [the Canadian government] can use is conservation. We have to look at the conservation of this stock, and we have to control them, we have to turn around and regulate it and stuff like

that. Where we already had our own conservation [plan] and we were
following it.

Cindy is suggesting that the government is using conservation to mask
a deeper interest in maintaining the status quo. The primary concern of
the government, in Cindy's eyes, was not that the fish stocks be successfully conserved, but that they remain in the control of federal agencies. But this interest in maintaining federal regulatory control is not
stated explicitly by the government, according to Cindy, which instead
uses the language of conservation to mask its position.

Lloyd Augustine, one of those who authored the *EFN Fisheries Management Plan* and keptin of the community, said that the plan itself was
not well received or officially acknowledged by the DFO, or by fisheries minister Dhaliwal.[10] In Lloyd's view, the federal government was in
a position where they could not acknowledge the productive conservationist stance of the plan because that would also require them to address the concerns for rights and sovereignty inherent in it. Instead, he
suggests, they avoided the question by pretending not to have received
the plan and then refusing to respond to it or to come to the community and meet with the plan's authors. If, as the community suggested
in the management plan, the federal government was using conservation as a political tool against the Mi'kmaq, then the government also
could not acknowledge the conservationist position of the community
as legitimate. Instead, in the eyes of many in Esgenoôpetitj, the federal
government used the rhetoric of conservation to justify the social controls being exercised upon the community. In order to maintain its
power, the government used increased amounts of policing pressure
– initially overtly and later through covert surveillance, threat, and
manipulation.[11]

Within both Burnt Churches, most people believe that the government and police were not motivated to do what was best for their communities. In English Burnt Church, the government's focus on control
is a vexing problem. People see the Canadian government as their own
and many are seeking ways to resolve the increasing distance they feel
from the decisions and decision makers that affect their lives. For
Esgenoôpetitj, where people are much less likely to see government
agencies as theirs (even if they do see themselves as Canadians), the
refusal of the DFO to recognize their conservation plans was a further
confirmation of the intent of the government to maintain and consoli-
date its power over Mi'kmaw people.

Conservation is a critical language for the negotiation of power throughout the dispute, precisely because it stands in for so many things, eliding values and concerns, rendering all political positions more approachable or appealing. Local concerns about livelihood, sovereignty, and place are not privileged in public conversations about the dispute, where people's ties to place and history, to local belief and culture, are poorly treated, if at all. The government does not appear to engage with claims at these deeper levels, insisting that it must focus on conservation and on law and order. Conservation then becomes a key issue and framework in the dispute: locals both use and critique the government's conservation discourse in an attempt to remake conservation as something that reflects their own values and concerns.

Seeking Allies, Practising Resistance

People in both Burnt Churches turn to the rhetoric of conservation as a way to get their voices heard in the public discourse and to mobilize support from allies outside of the immediate conflict. The articulation of conservationist positions does not mean that people are adopting the positions of globalized conservation groups such as Greenpeace, the Suzuki Foundation, or the World Wildlife Fund. The "saving nature for nature's sake" positions of the global conservation movement do not reflect enough of the concerns of residents of the two Burnt Churches. Rather, as I have demonstrated in the discussion of livelihood and government critique, the framework and language of conservation is adopted by people as a way to present their concerns in terms that are accessible to and acceptable in the broad public discourse. Talk of conservation in the dispute is important not only to rebuke the government and to argue for livelihood; the language of conservation itself is an important tool that people use to mobilize support for themselves and their community, both during the dispute and in its aftermath.

In both communities after the dispute, and especially in the English village, conservation remains an important way to describe community concerns and motivations, as people attempt to describe their roles in the dispute so that they might be received favourably by outsiders. During the course of my fieldwork, for example, conservation talk was one of the strategies people used to try and mobilize my support and empathy for their concerns. In the English community, fishers in particular talked at great length about the numbers of lobster traps in the water during the native fishery, the resulting millions of pounds of lobster caught, and the threat that these posed to the ongoing sustainability of the region. They described their own efforts to manage the fishery sustainably and the possible long-term effects of the entry of so many new fishers into the commercial fishery at the end of the dispute. Certainly, these concerns *are* at the crux of the matter for English fishers, especially as they relate to livelihood. Offering them up in the language of conservation is a way to seek solidarity with the listener in the space between saving fish for fish's sake and for fishers' sake. In the Mi'kmaw community, "conservation talk" is employed alongside "rights talk" in order to find allies. In many conversations, people used rights talk to frame the entry of natives into the fishery, and conservation talk as a way to further justify self-regulation of the fishery. While they did not frame conservation as the primary impetus of the dispute, it forms a key part of the community discourse as a response to critiques of the native fishery by government and commercial fishers. In the aftermath of the dispute, conservation remains an important tool in attempts to mobilize the support of outsiders, such as myself.

During the dispute, in the English village people felt they had no allies and no empathy from the "outside world." After the initial confrontation during the protest on the wharf in October 1999, few in the region, including other commercial fishers, came to the English village. Luke highlighted the impact of this isolation: "When all this racket was on us, after the cutting was all over, nobody showed up to support, to help or anything like that, we were stuck, this community all by itself ... It made it hard on this community." As an outsider studying the dispute, I came to Burnt Church with the impression that alliances were formed between English and Acadian fishers during the dispute. I found quickly that this was not so, at least in the experience of the English community. In every case, when I asked about solidarity among local settler groups over the course of the dispute, English residents said that it did not happen. The explosive situation in Burnt Church was not something that others wanted to involve themselves in. In the region and across the nation, those who involved themselves in the dispute were largely those motivated by solidarity with native fishers and their community, such as Christian Peacemaker Teams (CPT), the Aboriginal Rights Coalition-Atlantic (ARC-A), and the Warrior Societies. Conservation talk allows English residents to explain the concerns they were expressing during that Sunday protest and afterwards in a way that resonates with larger public concerns. It plays upon and furthers the government critique of the native fishery, arguing that the impact of fishing during the dispute was even higher than the government claimed. Conservation talk is also a way for the residents of English Burnt Church to renew focus on fishers and the fishery rather than on violence, allegations of racism, or questions of the legitimacy of their settlement.

Conservation talk in the English community is accompanied by a studied silence on the question of poaching. The framework of poaching was not used by non-natives to characterize the activities of native fishers during the dispute because the entire native fishery was understood by them to be "illegal." Now that natives are a part of the commercial fishery, poaching practices (the sale or consumption of undersized lobster) are equally a concern on-reserve and off-reserve. In Nova Scotia, McMullan and Perrier argue that poaching is "a routine form of everyday resistance" in rural fishing communities, in which those involved in traditional communal practices of fishing resist the government controlled regulation of the fishery (1997, 29–30). The introduction of government regulation of the commons was intended to protect communal fishery resources, but instead "exacerbated the rise of rapacious fishing" (57). In their analysis, while some forms of poaching are motivated by commerce, in rural Maritime communities some forms of poaching are important forms of social resistance.[12] In communities where basic loyalties remain with the fishing team and others in a shared harbour, some "poaching is a collective action in its own right and, like demonstrations, occupations and other political mobilizations, it preserves on a daily basis the struggle of the commons" (57). McMullan and Perrier argue convincingly that in the Maritimes "much poaching … is·an integral part of community life" (55), an important form of rural resistance. The silence within the English community on this subject reinforces the idea that positioning their fishery practices as conservation-minded is perceived as highly important. Poaching did come up when discussing the actions of Acadian commercial fishers in the region, some groups of whom are seen as poaching to excess. In these conversations, poaching was clearly positioned as something that others did, not fishers from the Burnt Church wharf. In another example, a local man did tell me that he *used to* go out and catch some lobster for food without a licence, but not anymore. Perhaps the experience of the dispute has shifted local practice, minimizing poaching because of conservation concerns. Or perhaps the experience of the dispute has simply silenced conversation about these ongoing forms of resistance, easily possible in a context where "silence and secrecy are the preferred methods of coping with trouble" (McMullan and Perrier 1997, 46). In either case, it is clear that people do not want to risk discussing subjects that might call their conservationist positions into question.

For many people within the Esgenoôpetitj First Nation, the conservation discourse of the dispute represented both an effort by the federal government to discredit the native fishery and an opportunity for natives to find allies in non-native Canada. The position of the federal government, from the community's perspective, was not simply that conservation principles were absent in the native fishery, as discussed earlier; the federal government also characterized the native fishery as harmful and damaging, operating with blatant disregard for the ongoing health of the resource. The government's conservation rhetoric was seen by Leo and many others as a form of propaganda war being waged against the community:

So the government came in afterwards and they applied and they used propaganda, everything against us to rile up the communities around us. Inaccurate numbers were one of [their tactics], talking about other [native] communities coming down and fishing and stuff like that, so all the numbers that they have was inaccurate and they put it out in the papers. 6000, 7000 traps, whatever, and that wasn't so.

In this view, the intent and effect of the government's tactics were to minimize the amount of support that the native fishers were getting from non-native sympathizers, particularly from environmental and social-justice activists, and to discredit the ability of the Mi'kmaq to regulate their own fishery. The government was seen to be allied with the fishing industry against the natives.

In response, the Mi'kmaq began to articulate their own conservation policies and attitudes publicly and explicitly through the management plans (*The Esgenoôpetitj Fishery Act* and *Draft for EFN Management Plan*) discussed in detail earlier. The specific conservation principles articulated within the plans became a very important foundation for alliances that were sought in the region. After the release of the management plan, a large group of regional scientists, environmentalists, and activists, including the Conservation Council of New Brunswick, endorsed it and its conservation principles. This was a significant victory for the people of the First Nation in their search for allies. Alana reflected on the challenges of offering a credible alternative to the government's position:

> The management plan that we had worked on was great. It was more geared towards the environment than the government's program ever thought of being. Environmentalists and scientists and different people coming in and looking at the plan and saying "This is great, this is wonderful ..." There was not going to be a problem. The government made it into a problem because, I don't know – greed? And because they just wanted to control [the fishery]. I think that that's the main thing.

Conservation talk, in the First Nation, was an important tool in positioning the native fishery in a favourable light, and in discrediting the government's position. Post-dispute, the conservationist nature of the community's position during the dispute remains something that uniformly articulates a part of the story of the conflict. Conservation talk allowed the people of Esgenoôpetitj/Burnt Church to address the government in the government's language, once it became apparent that the concerns of the people themselves were largely absent from federal rhetoric. If the rhetoric of the dispute was, in part, a propaganda war with the government over conservation, as was earlier suggested, the *Fishery Act* and *Management Plan* served the residents of Esgenoôpetitj as significant weapons.

Community Difference and Common Concern

Historically, the two communities of Burnt Church are separate places, divided along racial and geographic lines reinforced by the policies and practices of the Canadian colonial government. The ancestors of the English villagers of Burnt Church were among the agents of colonialism in the region, transforming part of this place into the rural anglophone fishing community that their descendants now know as home. Though the two communities share a name and a common landscape, they are in many ways very separate places. In contemporary times, the separation between these places leaves the communities unable to recognize their common interests or work together for common goals, because they continue to see one another as rivals for resources, political voice, and economic opportunity. This division was seen to be exploited by the Canadian government during the dispute as a way to buttress its own power and control in the situation. This government strategy will be explored in greater detail in the following chapter. During and after the dispute, the Burnt Church/Esgenoôpetitj First Nation and the English village of Burnt Church found their values and interests marginalized by the federal government, who also used conservation talk to maintain the separation between the two groups.

In Esgenoôpetitj, the profound legacy of colonialism over many centuries leaves people familiar with the experience of marginalization and motivated to resist the imposition of external powers and definitions on their community. Those most committed to sovereignty were also those providing leadership to the community during the dispute. The discourse of conservation became, for this community, a tool to resist the government's positions and to articulate their own values in a way that might be heard by the government and by its NGO-sector critics. They used conservation to set up a difference between themselves and the government and its commercial fishery, as practised by their English neighbours.

In the English village, the experience of marginalization from power by globalization and urbanization is a relatively new one. The perception that the status of their community with government and the Canadian public is slipping drives efforts to maintain social and political power with respect to Mi'kmaw neighbours and to government. Conservation talk becomes, for the English also, a tool to articulate local values within a framework that might be recognized by government and NGO-sector powers, as well as a way to try to position themselves in a more favourable light than their neighbours.

These similarities in values and in strategies between the two Burnt Churches are not recognized by the communities themselves. During and after the dispute, relationships of cooperation between the two groups are few and far between. In the early years of the dispute, the local United Church minister in the English community worked with some Mi'kmaw allies to hold community meetings between residents of these two places. They hoped that people could talk about their concerns and experiences, and perhaps find some common ground. For some of these meetings, the provincial government paid for a facilitator. But the conversations broke down. Many in the Mi'kmaw community felt that participating in such dialogue was probably not safe – personally, emotionally, and perhaps physically – and relied on a few community representatives to go into the English village for these meetings on their behalf. Within the English community, people felt that they were hearing the same old tales of pain and woe from the same people, stories that they could be responsible for. Over time, people from both groups stopped attending because they felt the meetings were not progressing in a productive manner. When the provincial government money for a facilitator was not renewed, the conversations stopped altogether. For moderates in the English community in particular, the failure of this effort demonstrates the profound difficulty of bringing the two communities together around shared concerns. Many people in each community continue to feel unsafe with their neighbours, never sure when racial violence will erupt.

In the meantime, a few individuals from both communities work to cross the divide, nurturing relationships with old acquaintances, attending community fundraisers, forging gentle ties. The organist at St Anne's parish, the Mi'kmaw congregation, comes every three weeks to St David's United Church in the English community to play for the service, as St David's is without a regular organist. One Sunday near Easter in the year when I was there, she played and sang an old hymn, *How Great Thou Art*, in both English and Mi'kmaw for the English congregation. While this seems like a small gesture, in this politicized and racialized environment it is, in fact, quite a risk, one that took clear commitment on the part of people from both communities. On the Burnt Church wharf, a very uneasy truce exists between non-native commercial fishers and native fishers who entered the fishery at the end of the dispute. These fishers find ways to work alongside one another in the day-to-day routines during the commercial season. The ongoing management of the wharf remains a profound challenge, though, as the communities continue to see each other more as rivals than as people with whom they share a common interest. Within the English community, those who hope for a better relationship between the two communities feel that it will come far in the future, with the next generations, based on threads of relationship nurtured in the present. Within the First Nation, people also hope that some change will come with future generations. But the legacy of conflict remains strong beside these slim hopes.

The next chapter explores some of the non-local senses of place at play in the dispute by considering the positions of the Canadian government and non-native Canadian activists in the dispute, and their implicit values and concerns. The extreme marginalization of indigenous peoples through colonization and the newer disregard for rural communities by globalized powers not only separate each of the Burnt

Church communities from decision makers and discourses of power, they also recreate and exploit the separation of these communities from one another. Reconciliation between neighbours depends on more than the work of local individuals: it also requires a transformation of the larger frameworks of power that rely upon their separation.

6 The Canadian Way

When one examines the dispute at Burnt Church Esgenoôpetitj, exploring each issue reveals underlying values which merit consideration. Though the dispute was concerned with conservation, it is clearly not simply about fish. For locals, concern for conservation is an expression of deep concern for their livelihoods and the future of their own communities, and is a way to seek allies and resist the controlling power of governmental practices. The dispute is also about rights – about the exercise of treaty rights among the Mi'kmaq at Esgenoôpetitj, and about the rights of the individual commercial fishers. But when we look closely at rights, we can see that rights are often frameworks to express deeper, implicit values and concerns, values which differ significantly between the two communities at Burnt Church. Among the English, rights are a way of talking about the individual's due, about the importance of equal treatment, about what should fall to every person by virtue of their being Canadian. For many Mi'kmaq, rights are a way of expressing the importance of the community as a whole, and of historical relationships. Rights are not simply about getting one's due, but a way to argue for the value of relationships as framed in the treaties to contemporary life. And what of place? Understanding that the dispute is about place means recognizing not just that the conflict is about material space, but that it operates on the level of values and world views.

While local constructions of place may be unique in that they are, to a significant extent, socially, historically, and geographically grounded, non-local senses of place also have important political and social power over specific places. As the geographers Williams and Stewart suggest, the social and historical processes that create place can involve non-local commodity interests, environmentalists, or recreation enthusiasts, for example: "Even what planners and scientists put forward as a data-driven description of a place in the form of a scientific assessment is itself another competing sense of that place" (1998, 20). In considering the challenges of managing competing senses of place for environmental planning and decision making, Williams and Stewart suggest that they all need to be understood as "legitimate, real, and strongly felt and an important source of political conflict" (20). In Burnt Church, local senses of place are the necessary starting point of our exploration of the dispute, yet many other senses of place operate there. Local values and priorities compete with those of government scientists, the media, RCMP and fisheries officers, members of Christian Peacemaker and Aboriginal Rights Coalition observer teams, Warriors, the Assembly of First Nations, the Conservation Council of New Brunswick, and the Maritime Fishermen's Union, to name but a few of the groups involved in the dispute over the years. Among the local people with whom I spoke, there was a profound awareness of the power of these external conceptions of place in influencing the dispute and its outcome. This chapter explores the role of the key non-local players in the dispute, of the Canadian government, and of its calculative approaches. It discusses the influence of nationalist myth making, and concludes with a discussion of the multiple values that motivated people in the dispute. I argue that recognizing and valuing a multiplicity of values, philosophies, and ways of Being is necessary to relationship building between indigenous and non-indigenous people in Canada.

Calculative and Canadian

The dispute in Burnt Church concerned a contested place, a place contested not only by local residents but by the Canadian government and Canadian citizens; as Williams and Stewart have described elsewhere, it was "a public exercise in describing, controlling and negotiating competing senses of place" (1998, 23). When place is contested, as it was in Burnt Church, local conflicts are embedded in larger networks of power. These powers shape local positions within the conflict, as was evident in the preceding chapter's discussion of conservation discourses; they are also players in the conflict themselves. During the fishing dispute in Burnt Church/Esgenoôpetitj, the Canadian government was a key player and party to the dispute. The government was clearly not prepared for the post-*Marshall* conflict. Their approach to the dispute was management focused and profoundly unable to acknowledge the diverse values motivating those involved. Instead, the Canadian government attempted to reduce the fishing dispute to questions of authority, regulation, and enforcement. This approach reflects the government's own sense of place, a calculative, data-driven, managerial view of Burnt Church and Esgenoôpetitj.

Fisheries management is a responsibility of the federal government, not of the provinces, and is carried out by the Department of Fisheries and Oceans (DFO). For most of the duration of the dispute, Herb Dhaliwal was the minister of fisheries and oceans (he was replaced by Robert Thibault in 2002). Aboriginal affairs are also a federal responsibility, falling to the Department of Indian Affairs and Northern Development, led by Minister Robert Nault. Finally, though some provinces have their own police forces, the province of New Brunswick is among those that contract their policing services to the federal police force, the Royal Canadian Mounted Police.[1] When the post-*Marshall* fishery began in Atlantic Canada, the government determined that the immediate issues to be resolved were those of fishery access and enforcement, which fell to Minister Dhaliwal and the DFO rather than to Indian Affairs.[2] The day-to-day participation of the Canadian government in the dispute was managed by bureaucrats in the DFO and carried out by DFO officers charged with enforcing fisheries regulations and RCMP officers charged with law enforcement. The work of enforcement was sometimes also supplemented by the officers and resources of the Canadian Coast Guard. Over time, both the RCMP and the DFO called in many additional officers to the Burnt Church region to help with enforcement in the dispute. Both agencies had regional offices in the Acadian community of Néguac, which served as bases for their operations. In addition, the RCMP set up a trailer "command post" at the occupied wharf in the English community of Burnt Church. The activities carried out there included surveillance of both communities, monitoring of protestors, policing of barricades, negotiating with protesters, and participating in enforcement activities on the waters. The DFO did not have a command post in the community; DFO officers arrived by boat on Miramichi Bay to survey the native fishery, seize native traps, and carry out other enforcement activities on the waters.

Admittedly, the government of Canada is not a monolith. It comprises various bureaucracies, political entities, and individual civil servants and officers holding different viewpoints and taking different approaches. In the Burnt Church dispute, there were more governmental actors involved than DFO and RCMP participants – such as local

Member of Parliament Charlie Hubbard (Liberal), the Coast Guard, the House of Commons Standing Committee on Fisheries and Oceans, the Department of Indian and Northern Affairs, and the federal cabinet. Yet, on the ground in the communities of Burnt Church, the two primary agencies of the government involved in the dispute, and which residents had to deal with, were the DFO and the RCMP. Post-dispute, there are faint suggestions that inter-agency relations between the DFO and RCMP were not always stellar. For example, one RCMP director writes, "RCMP relations with the DFO were challenged as the department's mandate was to protect the lobster fishery and enforce the Fisheries Act, whereas the RCMP's mandate was to prevent the loss of life and damage to property" (Vickers 2004, 3). In the local and public experiences of the dispute, such inter-agency conflicts were not obvious. For locals, the views and actions of these two parties taken together characterized the response of the Canadian government to the dispute; nationally, it was clear that the DFO and, to a lesser extent, the RCMP were responsible for the dispute on behalf of the government.

In her study of Mayan activism, *Indigenous Movements and Their Critics*, Kay Warren observes that "the experience of writing this book convinced me there is simply *no neutral* position or language of analysis through which to author the story of ethnic resurgence" (1998, xii; italics in original). This is also true in our discussion of events in Burnt Church; there is no neutral position from which to characterize the activities of the government (or of other players, including locals) in the dispute. In this project, I have taken the position that understanding the experiences and insights of local people in their own terms is critical. Certainly much more is left to be said about the complex story of the Canadian government's relationships with indigenous peoples, in Burnt Church and elsewhere, and about the complex negotiations that must have occurred within the government as it struggled to deal with the implications of the *Marshall* decision. I would like to have understood much more about the experiences of RCMP and DFO officers in the dispute and the internal politics of their bureaucracies. I was unable to do so – in the polarized post-dispute context, such research would have come at the expense of relationships with local people. My analysis of the Canadian government takes its cues both from local experiences and from the public self-representation of government agencies, specifically the RCMP and DFO.

The DFO argued that the conflict in Burnt Church was a stock management and access issue, and that they were the only legitimate authority who could regulate and enforce the fishery. In a statement at the beginning of the 2000 fishing season, Minister Dhaliwal argued that the authority to regulate the fishery was his alone, as fisheries minister:

> I as Minister have the authority and the responsibility to regulate the Aboriginal commercial fishery ... It appears that some, though not all, parties at Burnt Church want to regulate and control the fishery, independently of the Government of Canada.
>
>   One cannot assert only the part of a Supreme Court decision that one agrees with, and reject the rest. The fish resources are the common property of Canada; and the Supreme Court affirmed my authority and responsibility to regulate for conservation and other purposes. (DFO 2000a)

In the government's interpretation of the situation in Burnt Church, as conveyed by Dhaliwal, those who wish to challenge the ability of the Canadian government to regulate native fisheries are reading the decisions of the court selectively. Dhaliwal argues that all fish resources are Canada's, and that native access to the fishery is something the Canadian government is obliged to provide, albeit on its own terms. This view reduces the "legitimate issues" in the conflict to two, excluding historic concerns about place, sovereignty, and livelihood in both communities, and making the Burnt Church dispute one focused on government authority and stock management.

   In 2001, the DFO commissioned a fisheries scientist to study the lobster fishery in Miramichi Bay and to characterize the scientific and resource problems associated with it. The Caddy Report, as it was called, took twenty-five days to complete and was released by the DFO to demonstrate its ongoing concern for the lobster resource (Caddy 2001). Caddy's "data-driven" description of the lobster fishery in Miramichi Bay represents quite a different view of place than do the stories of locals. It is a powerful depiction of a place where stocks are in crisis and where, in order to resolve this problem, the biological understanding of the situation should take precedence. This view had great currency with the DFO, who in their press release on the report echoed Caddy's conclusions about the nature of the dispute: "A first conclusion from talking to those most involved ... is that there is an urgent need to raise the level of *public understanding of lobster biology*" (Caddy 2001, 13; cited in DFO 2001; italics added). This report buttresses the government's argument that the tensions in Burnt Church are about lobsters – who fishes lobsters and who decides who fishes lobsters. It continues to place stock management at the forefront of the dispute, framed in scientific and resource management terms. The problem is not understood to be the government's lack of appreciation for local concerns, but the public's lack of knowledge of lobster biology.

The federal government's lead agency in the dispute, the DFO, persisted in this reframing of the issues at the heart of the dispute in order to legitimize its authority and, perhaps, to ensure that it did not have to address the issues that motivated native fishers: rights, justice, and sovereignty (as explored in chapter 4). This occurred even when the government was ostensibly attending to community concerns and relationships, rather than the fishery itself. In January of 2002, as a part of its "*Marshall* Response Initiative," the DFO appointed Mr Justice Guy A. Richard and Chief Roger J. Augustine to head the Miramichi Bay Community Relations Panel.[3] The panel was charged to meet with local native and non-native community members in the Miramichi area and to assess the relationships between native and non-native communities, reporting back to the federal government with recommendations and approaches to improve relationships. Richard and Augustine spent many weeks meeting with locals in the area immediate to the dispute and across the region. Their report concludes that "the problem [in the dispute] runs much deeper than lobster fishing and conservation" (Augustine and Richard 2002a, 1–2). They recognize the importance of livelihood in non-native communities and of political autonomy and sovereignty to the Mi'kmaq of Burnt Church. And yet, their report does not deal with treaty rights, Aboriginal title, or sovereignty as issues in the dispute, as "the Panel was expressly precluded from addressing the question of Treaty Rights" in their mandate, which had been set by the DFO (3). Even when the government broadened its view of the dispute from fishery regulation to include community concerns, as with this panel, it remained unwilling to examine some of the issues most important to those communities.

On paper and in the media, the DFO represented Burnt Church as a place where fishery access is a problem, a problem compounded by the unwillingness of some natives to recognize the authority of the government to create and control access. The government believes its authority has been legitimated by the Supreme Court in the *Marshall* decision, and this authority is reinforced through continual appeal to the court's decisions as impartial, authoritative, and final. This is true not only when the DFO is positioning itself against native fishers but also when it goes up against right-wing critics within Canada, such as those within the former Alliance Party. In response to Alliance criticisms of the DFO's post-*Marshall* policies, Minister Dhaliwal argued repeatedly that the actions of his department were precisely in accordance with the Supreme Court's decision (and clarification) on *Marshall*, upholding his responsibility to hold an "orderly and regulated fishery," where any limitation on the constitutional rights of Aboriginal people "had to be justified on conservation or other valid public policy grounds" (DFO 2000b).

The legitimacy of the government's authority in the lobster fishery is challenged by native claims to sovereignty and by other readings of the treaties and of the *Marshall* decision, as has been demonstrated in chapters 4 and 5. The government refused to address these alternative arguments about authority and resource management, unwilling to entertain conversations not only about the political position the government took after *Marshall*, but also about its enforcement actions on the waters of Miramichi Bay and in the communities of Burnt Church/Esgenoôpetitj. The actions of the RCMP and the DFO to uphold the government's authority and an "orderly" fishery in Burnt Church included seizure of traps by RCMP officers in riot gear with assault rifles; the monitoring of Mi'kmaw residents with direct and electronic surveillance; chasing, swamping, and ramming native fishing boats; and the violent arrest of native protestors. These actions have been documented in the Canadian media, in the stories of locals, such as those written here, and in the report of Christian Peacemaker Teams (CPT) who observed the fishery. The extremity of the government's actions in the name of authority and order is demonstrated in the CPT account of the arrest of Mi'kmaw fisher Brian Bartibogue:

> August 13 – After a six week pause for the lobsters' moulting season, EFN [Esgenoôpetitj First Nation] fishers began to set traps again on August 10. At 11 p.m. on August 13, fourteen DFO boats arrived in the dark with no navigation lights and began seizing EFN traps. Two EFN boats approached, with CPT observer Nina Bailey-Dick on board one boat. DFO officers pointed their guns at the unarmed fishers in the other EFN boat and said, "Get back to shore or we'll shoot." A few minutes later another DFO boat rammed EFN band councillor Brian Bartibogue's fishing boat, arrested him and three others in the water, and confiscated his boat. Bartibogue was beaten and choked unconscious by DFO officers before being taken with the others to the Tracadie RCMP post. For several hours,

the RCMP denied the prisoners medical attention, dry clothing, and phone access, and lied about these conditions when observers Nina Bailey-Dick (CPT) and Ron Kelly (ARC) inquired about the prisoners' well-being. (Christian Peacemaker Teams 2001, 8)

In local and regional accounts of this story, it is said that Mr Bartibogue only received medical attention when he used his one phone call to dial 911 and request an ambulance, which attended him at the RCMP detachment. In the 2000 fishing season, the CPT report alleges, there were twenty-two incidents of government violation of the human rights of native fishers during their enforcement actions (7). The realities of the government's actions in the native fishery belie the claims of orderly regulation so common in the government's rhetoric.

In both Burnt Churches, residents encountered tension between the interests and attitudes of some individual officers and agents of the government, and the government writ large. Within the English community, people viewed the government's enforcement actions as inconsistent and inadequate, as they did not halt the native fishery. Because the English villagers were not seen to be fishing outside of regulations, they were more likely to have civil conversations with the RCMP and DFO officers in their community. They often heard from the officers that their superiors were unwilling to allow them to enforce the full extent of the law in the native fishery. They were told that the RCMP would be unable to protect them adequately from violent natives, and were asked to leave their homes for their own safety. It seems that individual officers were, or wanted to present themselves as, more sympathetic to the English community than their superiors. For most English residents, while they got information from these sympathetic officers, they had little sense that these individual positions had any effect on the positions of either the RCMP or the DFO.

In the First Nation, many people observed that one of the government's strategies as the dispute wore on was to call native police and fisheries officers into the area to assist with enforcement in the dispute. Though perhaps the government believed native officers would be seen as more trustworthy by the Mi'kmaq, the reality was far more complex. For example, in the account of Brian Bartibogue's arrest and beating told to me by his brother, Mr Bartibogue surrendered to the DFO because of the assurances of a native fisheries officer that he would not be hurt, when he was in fact "beaten and choked to unconsciousness" (Christian Peacemaker Teams 2001, 8). More often, indigenous officers were often seen as conflicted figures who were being asked to take sides against their own best interests.

During the dispute, the DFO and the RCMP characterized Burnt Church as a lawless place, where the future of the lobster fishery was being put at risk because of natives' disregard for the authority of the government to regulate and control fisheries. The consistent framing of the dispute in these terms permitted the government to avoid addressing the deeper concerns of local people involved in the dispute. This sense of place – grounded in concerns for authority, regulation, and "data-driven" descriptions of lobster stocks – drove the events of the dispute and limited possible outcomes. The government wanted to assert and to assure its own authority over this place, over these people and their fisheries. Some might argue that this is appropriate, because it is only the government that can have objective or neutral analysis regarding the fishery. But the government is not a neutral party in this dispute – it was signatory to the original treaties, agent of settlement and creator of the reserve system, prosecutor of Donald Marshall Jr, and participant in the violence at Burnt Church/Esgenoôpetitj. The government's laws are not impartial or disinterested; they are written to enforce norms and regulate behaviours, to codify values.

The values that underlie the government's actions cry out for closer examination. The government is acting to protect its ability to manage the fishery, to "plan, research, and organize"[4] human harvest of fish. In order to do this effectively, the government must work from an uncontested set of presumptions about ownership of resources, the nature of scientific knowledge, and the legitimacy of its own rule. To do otherwise would prevent it from acting efficiently – or, at least, with a view to efficiency. Heidegger calls this way of thinking calculative thinking. "In calculation, one studies, organizes, and computes explicitly given, empirical realities without pausing to inquire originatively about the essential meanings that sustain these investigations" (Stefanovic 2000, 23). The overarching aim of calculative practices (technologies) becomes "to regulate and secure the natural world" (27). "The world now appears as an object ... Nature becomes a gigantic gasoline station, an energy source for modern technology and industry" (Heidegger 1959, 50). This mode of thinking operates precisely so that the essential structures of its authority remain unquestioned. In Burnt Church, for local people both English and Mi'kmaw, the fundamental questions of legitimacy, belonging, and self-determination in their own place, of being-at-home, were critically important. In ignoring these questions, one might argue that the government tended to turn people and communities primarily into objects within its calculus, into instruments of the government's agenda. People were no longer ends in themselves but, principally, a means to realize the government's purpose. For many Mi'kmaq this was a familiar relationship to government, but for many among the English settler community, this was a shocking new experience. The Canadian government, never questioning its own assumptions, was seen as refusing to recognize its own role and position as a party to the dispute.

Canadian Myth

Nationalist myths convey powerful non-local senses of place and can become tools used to justify the necessity of the calculative regulation and management of specific places by external authorities. In many conflicts over place, nationalist myths are invoked to legitimate the authority of governments to regulate and protect nature in ways that often exclude the interests of local people. Local places are stripped of their human history, framed as "natural" places, and linked to "a state myth which legitimates protectionist action" (Berglund and Anderson 2003, 5). Berglund and Anderson suggest that "any effort to save 'it' [nature] then is linked to the political question of who should manage it ... Nature becomes the province of experts regardless of who occupies it and, furthermore, provides grounds for discriminating against the very people who do" (5). In this context, issues of authority become the central concern of ecopolitics, and local people and their conceptions of place become obscured by conflicts over the power to regulate and control. The classic North American example of this myth is the myth of the frontier, which "enabled white colonizers to justify the dispossession and slaughter of indigenous populations" and which is also "the founding myth of American environmentalism" (5). These nationalist, naturalist myths also bind Canadian identity together – culturally, historically, and communally – in a civil religion. The local power of civil religion was illustrated in the discussion of Canada Day celebrations and the installation of the Burnt Church cenotaph in the English village at the end of chapter 4. At the national level, these myths legitimize the construction and imposition of power relationships, both for the Canadian government and also for citizen activists working in opposition to the government.

In the Burnt Church dispute, we have seen how the DFO appeals to a national institution, the Supreme Court of Canada, which it portrays as impartial arbiter and ultimate authority. Further, the DFO characterizes its actions as maintaining or bringing "order" to the fishery. This appeal to "order" echoes the 1867 Constitution Act, in which the phrase "peace, order and good government" is used in granting the highest authority to the federal government (Bélanger 2001). Over time in Canada, the phrase "peace, order and good government" has become a part of the language of national myth, familiar in the discourse of Canadian identity.[5] In the Burnt Church dispute, the government's appeal to order might be understood as an effort to uphold its own interpretation of the conflict in the language of "historic" Canadian values.

The most fascinating example of an appeal to national myth in the dispute appeared in 2004 in an article in the RCMP *Gazette* by Kevin Vickers, "The RCMP and the Canadian Way: Using Lessons from the Past to Build a Modern Policing Philosophy." Vickers was the officer in charge of the RCMP at Burnt Church during the dispute. He invokes the story of Sitting Bull and the Sioux people's refuge in Canada after their victory in the Battle of Little Bighorn as a model for Canadian policing. He argues that by finding a way to allow Sitting Bull and his people to remain peacefully in Canada, the RCMP were laying the foundation of the "Canadian Way," a distinctly Canadian policing philosophy. The "Canadian Way" is "not founded solely on the rule of law, but rather on respect of human dignity. The Canadian Way is one of creative problem solving. Our approach includes respect, dialogue, facilitation, empathy, education, and, when necessary, enforcement" (Vickers 2004, 1). In this light, Vickers characterizes the efforts of the RCMP in Burnt Church as epitomizing this "Canadian Way," balancing the competing pressures and opinions of media, other government agencies, and local non-natives as they built relationship with the Mi'kmaw protesters. When a native barricade was erected along the main highway through the reserve, he suggests that, "instead of a confrontation, RCMP members, acting as facilitators, showed up with coffee and doughnuts to begin dialogue with those manning the barricades" (2). The more direct actions of the RCMP on the waters, he suggests, were not a part of this "Canadian Way," but necessary as a result of the RCMP's obligation to assist other federal agencies, such as the DFO, in their mandates. In the end, Vickers characterizes the response of the RCMP in Burnt Church as directly descending from the "example of communication and respect that was set back in the 1860s" with Sitting Bull, policing the "Canadian Way" (3).

The contrast between Vickers's characterization of his force's actions here and the stories of local people in previous chapters is very stark.

Though there were occasions when people in each community felt positively about the presence of the RCMP, for the most part their actions were seen as oppressive and violent in the First Nation and inadequate or unpredictable in the English village. It is also important to note that even the mythic events of the refuge of Sitting Bull and his people in Canada did not end well for the Sioux. They had no access to food or other resources in Canada and the government denied their requests for a reserve. They were essentially starved out, forced to return to the United States, where Sitting Bull and his family were held as prisoners of war for two years. Sitting Bull made a living as a "Show Indian" in Buffalo Bill Cody's Wild West Show, before he was killed while being arrested on the orders of a U.S. Indian agent. To legitimize his force's actions in the dispute, Vickers appeals directly to nationalist myth, invoking the "Canadian Way" to frame the activities of the RCMP in Burnt Church as non-violent and collaborative. Yet neither the myth nor the Burnt Church dispute were actually non-violent or collaborative, and neither ended in outcomes that address the real concerns of First Nations people.

For the people of the Burnt Church First Nation and their sympathizers, the actions of the Canadian government through the DFO and the RCMP were a terrible injustice, which, if recognized by the Canadian public, would have destabilized the government's position in the dispute. Addressing the concerns of the sovereigntists would have required the Canadian government to address its own complex and difficult colonial history, in which questions about the legitimacy of Canada's displacement of Aboriginal people are reasonable and important. Instead, throughout the dispute and its aftermath, the government's agencies argued for their own views of the dispute, grounding those views in nationalist language and myth in order to invoke the propriety of their position. Acknowledging indigenous and local senses of place as reasonably belonging in the conversation about Burnt Church could undermine not only the government's authority but also its legitimacy; and so, for the government, this had to be avoided at (almost) all costs.

Non-native Canadian citizens were also involved in the dispute, as activists and observers in solidarity with the First Nation. These people articulated an alternative understanding of events in Burnt Church sympathetic to native rights and critical of the federal government's actions, characterizing the government itself as one of the key players in the conflict rather than an impartial arbiter. The majority of these activists were members of two groups, the Christian Peacemaker Teams (CPT), an international violence-reduction program of the Mennonite and Quaker churches, and the Aboriginal Rights Coalition–Atlantic (ARC-A), a regional coalition of United, Mennonite, Roman Catholic, and Quaker churches interested individuals and others. At the invitation of the First Nation, these groups sent trained observer teams into Esgenoôpetitj who monitored the fishery from shore and accompanied natives in their boats, documenting the government's enforcement activities on film and video. Their work was supported through donations from across their coalitions' networks, in the Atlantic region and nationally. Their accounts of the dispute, made public in reports, articles, and press releases, were attempts both to influence the public perception of the dispute and to put pressure on the Canadian government. They attempted to challenge many of the myths Canadians have about themselves and their country by witnessing and exposing what they see as the injustice in the government's actions in Burnt Church and presenting this injustice within the larger framework of Canadian injustice against First Nations people.

Interestingly, in an online diary chronicling his solidarity work in Burnt Church for the CBC, CPT member John Finlay not only critiques some myths of Canadian nationalism and identity, but invokes others to support his position, using the language of the Canadian national anthem: "Much of our time here in Burnt Church is spent in true Canadian fashion (as in our national anthem) in that we are "on guard" for up to twenty hours each day. In this case that means sitting along the shore with binoculars, a cell phone, and some photography equipment maintaining a watch of the bay" (2000, 15). In this case, Finlay is suggesting that the work of CPT is essentially Canadian in the sense that they (Canadians) are standing "on guard" for justice, against their government, in solidarity with the Mi'kmaq. In order to frame his own actions in positive terms – and win some sympathy for his justice position from the Canadian public – Finlay invokes another myth of Canada, a metaphor to compete with the government's metaphor.

Within the activist circles described here, there was a profound effort to take seriously the concerns of Aboriginal people and the values and experiences that motivated them in the dispute. For these activists, the dispute was about race, justice, and colonial history. Their involvement was an attempt to right ongoing wrongs, by standing with the Mi'kmaq against the federal government. Both the Mi'kmaq and their allies have mixed feelings about the outcome of the dispute, since it increased their access to the fishery without addressing their deeper concerns. Many took great risks to stand with Mi'kmaw fishers against the Canadian government in what they believed was justice and solidarity. From the perspective of the native community, though the presence of these activists in their community did not come without its challenges (particularly in building trust), some believed that the activist presence tempered the government's enforcement actions and even saved lives.

In the English community, the presence of the peacemakers and observers was galling, precisely because they took the Aboriginal view so seriously and apparently had so little understanding of other local experiences. Mary argues that the outside observers were not at all interested in helping non-native Burnt Church:

> They were there to help the Indians – they weren't there to try and make peace between the two communities ... They were there to promote the Indians' thinking with the government, and to put us down because we weren't doing what they thought we should be doing.

Generally, Canadian activists who involved themselves in the dispute did so with great sympathy for Aboriginal people involved in the fishery and for their sense of place, and with much less sympathy for non-native locals. John Finlay's online diary reflects this challenge: "It is really difficult to get a true sense of the depth of the feelings which the non-native fishers have. They obviously perceive a threat to their livelihood and way of life, but what else is it based on in addition to the negative feelings towards Mi'kmaq which have been learned at a very early age?" (Finlay 2000, 12). In the "activist view" characterized here, the primary conflict at Burnt Church was between the lawful native fishery and the unlawful government. Non-native residents are seen to be peripheral to the situation since the conflict with the government is not "theirs"; and they are also seen to be unsympathetic since their concerns are so often positioned as opposing native claims.[6] The activist discourse for Aboriginal justice replicates some of the faults Guha (1989a) and Vandergeest and DuPuis (1996) have described in the discourse of conservation, marginalizing the values of rural (non-native) people and attempting to impose the values of the activist justice discourse upon them. Mi'kmaw sense of place, the ties of the people to the land and waters of Burnt Church, are appropriately and importantly recognized. Because they do not conform with the views of non-native activists, the perspectives of non-native locals find no place – and seem, to many of these Canadian activists, neither penetrable nor truly relevant.

For Canadians, reflection and analysis about their role as settlers in a colonial nation – and about their relationship to place in this context – is a fundamental challenge. The liberal discourse of equality often denies that racism is a systemic or everyday problem in Canadian society, promoting instead a "'national story' of benevolence and generosity" (Srivastava 2005, 35). Srivastava suggests that Canadians operate within "contemporary national discourses of tolerance, multiculturalism and nonracism" that mask ongoing racialized conflicts (35). Addressing the racialized structure of society is profoundly challenging because Canadian moral identity is so tied up in a vision of equality, a vision that, like all national visions, "requires not only sameness and communion but also forgetting difference and oppression" (39). This vision of sameness and non-racism is fundamental to the vision of Canada which the government sought to uphold in Burnt Church.

Confronting the racism inherent in Canada's relationship with Aboriginal peoples requires confronting fundamental questions about Canada's history and legitimacy as a colonial state. Taiaiake Alfred, an indigenist academic, argues that

> most Settlers are in denial. They know that the foundations of their countries are corrupt, and they know that their countries are "colonial" in historical terms, but they still refuse to see and accept the fact that there can be no rhetorical transcendence and retelling of the past to make it right without making fundamental changes to their government, society, and the way they live ... To deny the truth is an essential cultural and psychological process in Settler society. (Srivastava 2005, 107)

Like the English residents, many settlers know Canada as their only home and wonder, as some of those I interviewed in Burnt Church did, why they must pay for the sins of their forefathers. But the problems inherent in Canadians' relationships with indigenous peoples are not only historical; they exist in individual, social, and political lives in the present. The fundamental discomfort of reflection on race and racism makes it difficult for many Canadians to reflect upon their shared position, with the English villagers, in the colonial present. Addressing the difficult relationships between non-indigenous and indigenous peoples in Canada requires thorough examination of Canadian experiences and institutions. This is not a call for some kind of collective Canadian self-abasement and emotivism, or an admonishment to "repent" the evils of personal and institutional racism; floods of guilt and sorrow only become another barrier to change. It *is* a call for more practical engagement with the real challenges of relationships between peoples of many nations and communities who inhabit the same places.

Perhaps it is not possible to engage in a discussion of place without invoking mythic notions – as place itself is so deeply felt, involves rational and non-rational forms of knowledge, is the necessary ground of experience. One member of the Christian Peacemaker Teams described his work in Burnt Church using the language of the Canadian national anthem: as being "on guard" (Finlay 2000). For Canadians, the problem at the heart of such disputes, echoed in another anthemic trope, is that these places are both "our home *and* native land." Addressing disputes such as the one in Burnt Church/Esgenoôpetitj requires that indigenous and settler senses of place are taken seriously, in their own terms, acknowledging the multiplicity inherent in colonial relationships.

What, Then, Was the Dispute *Really* About?

The dispute subsided in 2002, when the federal government and the First Nation agreed to terms under which the Mi'kmaq would recognize DFO regulation of the fishery. The questions of fisheries management were, for the moment, resolved. In the two communities of Burnt Church, while people were relieved that the violence was over, few felt that the conflict itself was resolved, or that their own concerns had been addressed. Coming to an agreement that deals with fishery management, and which makes provisions for Mi'kmaq access to a commercial fishery under the same regulations as other commercial fishers, resolves one issue. But the dispute was a far more complicated situation, and many values remain unacknowledged and unsatisfied. While the dispute was "about fish," it was also about the many other concerns, the values, hopes, and histories which I have explored here.

After the dispute, English villagers remain profoundly unsettled and uneasy, careful in their newly realized vulnerability. For them, while the dispute was about fish, it was also about their rights, livelihoods, and the economic future of their community; about their legitimacy and sense of being-at-home in a contested place; about their identity among their neighbours, and as members of the Canadian nation state. Conservation of the fishery was their central concern, not as a concern for fish, but a concern for the fishery as *theirs*, in all of the ways I've listed here. The Mi'kmaq were also concerned with the fishery as theirs, but what this means in the Mi'kmaw context is quite distinct from what it means for their neighbours. The people of Esgenoôpetitj were concerned with upholding their treaty rights to fish, because of their needs for livelihood and economic opportunity, and also because some saw the fishery as an opportunity to reclaim a Mi'kmaw way of life. Mi'kmaw sovereignty, rights, and community identity were central to a renewed Mi'kmaq activism. This was, in the eyes of some, an opportunity for the Mi'kmaq, grounded in traditional ways, to reclaim their rightful identity as a Nation, and to re-establish their own senses of authority and belonging in their own territories. For the Mi'kmaq, this is what the dispute was about.

Local people were contesting issues which are profound in their nature. History, identity, relationship, belonging, and the ability to be-at-home in your own community – these are constitutive elements of place. Yet their ability to negotiate these concerns was arrested by a government which refused even to recognize the existence or importance of local values different from its own. For the federal government and its agents, the dispute was arguably only "about fish," and about the management authority of the government itself in this place. While local players articulated their diverse values connected to community life and to the fishery, the government appeared to persist in its narrow definition of the conflict and apparently refused to entertain any conversations which might call this definition into question. This is the hallmark of calculative thinking, thinking which focuses on solutions without ever considering whether it is asking the right questions.

The dispute is about myriad existential questions, which must be seen as embedded within the colonial aftermath. The historical and political processes of colonialism and settlement in Burnt Church have created two communities who live side by side in the same place, where the existence of each community threatens that of the other. The ancestral and historical relationships which have shaped Burnt Church leave people with terrible questions about their responsibility for their ancestor's actions, about their ability to live with the trauma of historical remembering, and about the possibility of a future in their own homes. These questions had been hidden for so long that some forgot that they were there, until the dispute thrust them into the light.

The government's attempts to close off these questions, and to deny their relevance and even their existence, are attempts to deny the reality of the colonial past, and of the historical present. The dispute became enmeshed in the government's inability to acknowledge the existential, historic, and colonial questions of the dispute, and its need to be the one dominant player who defines all terms, and all possible outcomes. These are the processes of governmentality, which serve to reinforce the power of the state and its "institutions, procedures, analyses and reflections, calculations, and tactics" (Foucault 2007, 108). The calculative paradigms which drive the government's sense of place in Burnt Church denied the multiple values, multiple communities, and multiple places which were at issue. This denial of the very possibility of difference is characteristic of the colonial project, which seeks to reconstitute local diversity as elements of the structures of the Nation. The persistent existence of difference remains the most fundamental challenge mounted against the legitimacy of such powers, who defend their own authority by denying difference.

This fundamental denial of multiple realities became the most powerfully unsettling factor in the conflict. The government persistently enforced the view that only one party can be right, and this party can be neither the English settlers nor the Mi'kmaq. The only possible winner becomes the compromised state, which has been fundamentally unable to recognize the concerns of its own people. The dispute is really about fish and about place, rights, livelihood, sovereignty, community, history, and conservation, all things which mean something very different for each of the Burnt Churches, and within each community. The dispute is also really about the suppression of these multiple concerns, in the service of the managerial and instrumental calculations of government.

The deepest tyranny of calculative thinking lies in its ability to perpetuate the notion that it is the only way of thinking, and the only possible way of moving forward. Calculative thinking has tremendous and effective value when it serves us along with other more originative modes of thought, when it is once again one among many. It is possible for originative and calculative thinking to coexist, even in relation to the concerns of the Canadian government. A stunning example of this is the work of Mr Justice Thomas Berger in the Mackenzie Valley Pipeline inquiry, which was carried out in the mid-1970s. To determine the appropriateness of building an oil pipeline through the traditional territories of Dene and Inuit people in Canada's North, Berger travelled extensively through northern communities, holding hearings to listen to the insights and concerns of everyone who wished to speak. His resulting report, *Northern Frontier, Northern Homeland* (1977) recommended a ten-year moratorium on development. More importantly, it took seriously the knowledge and values of local people, and used their insights alongside those of other more typical experts as a basis for his analysis of the pipeline. Berger's renowned report is an inspiring model of the depth of insight that is possible when multiple values are taken seriously, together.

The Peace and Friendship Treaties themselves serve as another model of the possibility and practice of pluralism in Canada. In 1761, the British and the Mi'kmaq signed the last in the series of treaties of peace and friendship in Mi'kma'ki. While the treaties have been until recently interpreted solely in their narrow written (English) form, these written treaties are now recognized as one element of a complex negotiation of multiple political and social relationships. In eighteenth century Mi'kma'ki, "societies that for the most part had been able to live separate from each other, could do so no longer, and were being forced to create a new political order that would make co-existence possible. In this sense, all of the British-Mi'kmaq treaties reflected a conscious decision by both parties to participate in the creation of a new political order" (Wicken 2002, 218–19). The treaties attempted to create a "new political order" in which British and Mi'kmaq ideas and practices coexisted in the same place. Our records of these agreements are thin, since the written English copies survived, but the wampum belts did not. Both written and oral accounts confirm that these treaties did not signify that the Mi'kmaq had given away their homelands, nor did they indicate the Mi'kmaq had unilaterally acquiesced to the British crown. They were an attempt at creating a new relationship between two nations in one place, the creation of a plural place.

The Peace and Friendship Treaties were the basis of Donald Marshall Jr's case before the Supreme Court of Canada in 1999. Wicken, whom I've quoted above, testified as a historian for Marshall and the Mi'kmaq. It was the court's decision to uphold the treaties, broadly understood in both oral and written context, which prompted the Mi'kmaw fishery in 1999, and the dispute in Burnt Church. Alfred argues that, in contemporary Canada, "the only possibility of a just relationship … [is] the kind of relationship reflected in the original treaties of peace and friendship consecrated between indigenous peoples and the newcomers when white people first started arriving in our territories" (2005, 156). Upholding the relationships framed in the treaties means renewing these relationships, not interpreting them within existing laws or management frameworks. Rather than trying to force a united sense of place in Burnt Church, or determine which group should prevail, we might instead look for ways to preserve the multiplicity that is always already there, and "learn to be pluralist" (266). These are the visions of the original treaties, which attempt to articulate the ways in which many different people might be-at-home together, in one place which is simultaneously many.

Postscript

Life in Burnt Church is not fixed; people's views, concerns, and relationships continue to shift and change over time. The stories and ideas discussed in this book reflect life in Burnt Church/Esgenoôpetitj in 2004–5. As the years pass, some things stay the same. Some things change. The joys of daily life in the Burnt Churches continue to develop, side by side with the challenges.

In the summers of 2007 and 2009, I returned to Burnt Church to visit friends and colleagues and to share excerpts and ideas from my writing about the dispute with those who were interested. In the First Nation, life continues to be challenging. Although more people have work in the fishery, poverty, overcrowding, addiction, and despair are on the resurgence, and some of those with whom I spoke feel that conditions in the reserve are declining again. The elected chief and council moved the fisheries offices and officers out of the new building built for their work into the old fire hall, and opened a small gambling establishment in the new building to generate revenue. The charismatic Bible study group continues to meet, though it is now led from within the Mi'kmaw community. In 2011 the elected chief, Wilbur Dedam, was removed from office by the federal Department of Indian Affairs along with three other councillors, including his wife Irene, on allegations of vote buying.

In the English community, the day-to-day rhythms and routines remain. Some men have left the community for employment in Alberta's oil fields because there is little new work to be had locally. There is a new minister at St David's United Church, but otherwise people's lives and occupations remain much as they were. At the Burnt Church wharf, two native boats currently join the non-native fishers for the commercial

Figure 7. As the Mi'kmaw fishery for food and ceremonial purposes begins in the fall of 2009, Leo's boat is among many being loaded with traps at the Burnt Church Wharf. This boat flies three flags: a Warrior flag, a Toronto Maple Leafs flag, and a skull and crossbones. (Photo: Sarah King)

season; the rest of the native fishers dock their boats in neighbouring Néguac or Tabusintac, or they fish from small boats and dories. The commercial lobster fishery continues to be in conflict; traps are cut when people from the "wrong" wharf are thought to be fishing in the "wrong" area. Catches seem to continue to decline, and some are very pessimistic about the future of the fishery. Lucrative though the lobster fishery may be in dollar terms, it remains a challenging industry, requiring increasingly intensive work for apparently diminishing catches.

In 2009 I visited Burnt Church/Esgnoôpetitj during the late summer. Mi'kmaw fishers were embarking upon the fall fishery from the Burnt Church wharf.[1] At Esgnoôpetitj, people were celebrating the annual powwow, where local lobster is the highlight of the final feast. In the

English village, community members put together a variety show of skits, jokes, and songs, as a fundraiser for St David's Church that packed the Women's Institute Hall. Others were busy writing and compiling a book about village history. Is the dispute now history for these people? Perhaps. The violence is over. But the values and needs of the people in this place remain potent, and unmet.

Notes

1. Introduction

1 This is the same Donald Marshall Jr who was wrongfully convicted of murder as a teenager. His eventual exoneration led to significant changes in the administration of criminal justice in Canada.

2 *R. v. Marshall*, Supreme Court of Canada (file 26014), 17 Sept. 1999, 2, http://www.lexum.umontreal.ca/csc-scc/en/pub/1999/vol3/html/1999scr3_0456.html.

3 Italicized quotes are taken directly from interviews carried out by the author in 2004 and 2005. The names Dalton and Cindy are pseudonyms; all pseudonyms are indicated as such when introduced.

4 This is a pseudonym.

5 The official lobster season in Burnt Church opens in spring and finishes in early summer. Since the Supreme Court's earlier *Sparrow* decision, a small native fishery has operated in the fall for "traditional and ceremonial purposes."

6 The traditional territory of the Mi'kmaq encompasses Atlantic Canada, the Gaspé Peninsula, and parts of Maine.

7 For specifics of the government's position, see chapter 2. For further discussion of its implications, see chapters 4 and 5.

8 Administered by the Unama'ki College at Cape Breton University; http://www.cbu.ca/unamaki/research.

9 Kovach is also among the many who argue that self-location is important in qualitative research (2009, 110–13). To locate myself, I should explain that I am a non-native Canadian, from southern Ontario, who grew up in rural Ontario; Fredericton, New Brunswick; and Toronto. Many of my ancestors arrived in Canada generations ago. I am among those who, if

I had to "go back where I came from," would not know where to go. Why am I writing about relationships between settlers and indigenous people? In part, because I've struggled with the difficult dynamics of these relationships and the ways they are represented. As a teenager, I was at the General Council of the United Church of Canada during the Oka crisis, and I heard elders and community leaders speak about their experiences behind the barricades. In such cases, it seemed to me that the public portrayal of these relationships did not reflect people's lived experience. Doing research in Burnt Church/Esgenoôpetitj allowed me the chance to explore this disjuncture.

10 This project was originally conceived as one that would include a wide variety of participants in the dispute at Burnt Church, including Acadian fishers and neighbours, for example. Once I arrived in Burnt Church, it became more and more clear that extensive ethnographic work was required to cultivate the depth of analysis intended. For this reason, the project was refocused on the places where the dispute happened, the two communities of Burnt Church.

2. "Those Relationships Became Countries"

1 During the year I spent in Burnt Church, many people initiated conversations with me about the dispute, wanting to hear about my research and to tell me about their own experiences. In the English village and in the First Nation, some people spoke to me about their experiences of the dispute the moment they heard of my interest and others never raised the subject. In Esgenoôpetitj, while some were very interested or at least willing to talk with me about the dispute, others were not at all interested in bringing up this history. Some were still traumatized by their experiences; some did not want to bring up a part of their community's life they did not feel supportive of; some, given their past experiences, were not willing to talk to an outsider they did not know. In my conversation with Miigam'agan we talked about these difficulties. "One day," she said, "I hope there will be a time when someone from within the community can gather these stories and write about our experiences."

In the English village, almost everyone I approached to participate in an interview agreed to do so. In fact, more people were interested in being interviewed about their experiences than I was able to speak with. Unlike their neighbours in Esgenoôpetitj, people in the village of Burnt Church have little experience of being poked, prodded, and misrepresented by anthropologists and researchers interested in their culture and world views. Like their neighbours in Esgenoôpetitj, they have much experience of being poked, prodded, and misrepresented by the media and by government agents. Once it was established that I was not a member of the media or working for the government, people were interested in building relationships with me, and our shared cultural positions made this easier than it might otherwise have been.

2   The majority held that these rights to fish for trading purposes were limited to those that would enable Marshall to earn a "moderate livelihood," and that they could be regulated by the minister of fisheries and oceans if this was done in a way that did not "infringe on his right to trade for sustenance" (R. v. Marshall 1999a, 3).

3   The first Peace and Friendship Treaty was negotiated between the Mi'kmaq and the British in 1725–6. This treaty became the basis for the subsequent peace and friendship treaties, the last of which was signed in 1761 (Wicken 2002, 3). For a thorough discussion of this treaty and its relevance to *Marshall*, see Wicken's *Mi'kmaq Treaties on Trial* (2002). Wicken articulates a broad view of treaties, emphasizing that a treaty should be seen as a written text that refers to an oral context – a context of relationship (12).

4   The grammar in this excerpt, and others from Miigam'agan, is her own, based upon her review of interview transcripts.

5   "Cutting traps" means either cutting the traps from their buoys so that they can't be retrieved, or hauling up the traps and cutting their nets out before throwing them back in so that they can't catch anything. The first method is ecologically damaging, as it creates "ghost traps" that sit on the bottom and catch lobster which can never be retrieved. Local fishers insist that any traps cut need to be done "properly," and not create more ghost traps.

6   This is a pseudonym.

7   This is a pseudonym.

8   The Mohawk flag mentioned here is probably the Warrior flag, which is used as a banner by Warrior Societies across Indian Country.

9   The provincial edition of the Saint John *Telegraph Journal*.

10 For discussion of this "clarification" see Wildsmith 2001, Wicken 2002, and Cameron 2009. Wicken characterizes the court as "reinventing what they had just said" (235).

11 James Ward (Sakej) was also one of the leaders of the Warrior Society at Burnt Church; Augustine is the traditional chief (keptin) of the area, as previously mentioned.

12 The story continues: "We went down to the court with him, and the judge said, 'You mean to tell me this one man beat up nine officers?' The nine

officers stand up – they're all big guys, eh? 'This one man beat all of you?'
And they said 'yes.' He [the judge] said, 'This is ridiculous!'"

13 See, for example, the CPT report *Gunboat Diplomacy: Canada's Abuse of Human Rights at Esgenoôpetitj (Burnt Church, New Brunswick)* (Christian Peacemaker Teams, 2001).

14 This is a pseudonym.

15 This is a pseudonym.

3. Contested Place

1 One of these groups was from Kairos, a Canadian ecumenical (Christian) organization interested in social justice. Kairos was hosting a visit from a Palestinian women's organization and arranged for a delegation of women to come to Burnt Church as part of their peace-building tour of Maritime Canada. Another was a group from the Katimavik program, a youth volunteer program of the Canadian government. A group of ten young adults were living in Miramichi and had an interest in the dispute, so I arranged a visit for them with some members of the First Nation.

2 It was probably partly due to the winter cold, or other considerations, but none of these tours actually stopped at the wharf. The only stop on the English side was at my home, an established "safer" space.

3 See chapter 4 for discussion of the Grand Council.

4 In this sense, this book can also be understood as a place-making project, as it constructs and reconstructs the history of the Burnt Churches over decades and centuries.

5 It was known most commonly as Pointe-à-l'Église, but various maps also show it as Chenabodiche, Pointe-à-la-Croix (Cross Point), and Pointe-du-Village (Village Point) (Basque 1991, 44).

6 Much of this early history, from the perspective of missionaries, was documented in the *Jesuit Relations*. The missionaries generally believed that they were in New France to save as many Indian souls as possible, whatever the cost, for the sake of the Indians and for their own sakes. Most were prepared to be martyrs to the faith (Greer 2000). It is more difficult to parse out how the Mi'kmaq saw the missionaries. Alliances and treaties between native nations and groups were commonplace on Turtle Island before the arrival of the colonists, however, so alliances with the French or British would fit easily into Mi'kmaw practices. In some cases, people from the allied nations would move to the other community, as a demonstration of commitment. Perhaps the baptisms of the Mi'kmaq and the presence of the missionaries in Mi'kma'ki were seen as dimensions of the relationship

between the Mi'kmaq and the French (see Prins 1996 on this point). And, as in other cases, it is difficult to know what Mi'kmaw converts thought and believed about Catholicism or about their traditional religions outside of what was written by the missionaries to document their work. For more about this, see work by Allan Greer (2000, 2005), Bruce Trigger (1985, 1996), Daniel Paul (2000), Harald Prins (1996), L.F.S. Upton (1979).

7 For a detailed account of this history around Burnt Church and Néguac, see Basque 1991.

8 For discussion of the larger biological impacts of European colonization on indigenous peoples in the Americas, see Crosby 1972, 1986.

9 The cenotaph recognizes the contributions of English and Mi'kmaw residents of Burnt Church. For more information, see chapter 3.

10 It certainly was not the case that every presenter was arguing this explicitly. Rather, it was implicit or explicit within some of the papers and within the questions and answers about the papers from the audience.

11 This is not an unreasonable fear, as any student of the East Coast fisheries knows, or anyone who has heard of the collapse of the cod fishery in Newfoundland understands.

12 This system is changing over time. As the population grows in the First Nation, new homes are built both by family members in family sections and also in newly cleared parts of the community where there are not sections. In the English community, population growth isn't a factor and so changes are slower, with the exception of the cottages along the shore. Many of these are still owned by descendents of old families, but many are not.

4. Seeking Justice

1 See Wicken 2002 for the texts of the 1726/6 and 1760/1 treaties, along with an extensive discussion of their historical and legal import.

2 This was known in both communities: while people in the First Nation were aware of the surveillance because they were subject to it, people in the English community were told of the surveillance by the local RCMP, who thought it would reassure them that they were being appropriately protected.

3 There are disagreements between people in the two communities over who had guns, and when and where. Certainly some people in each community *were* armed at different points during the dispute, according to stories people told me about their own actions. At the same time, there was also a movement within the First Nation to avoid the use of firearms.

4 See chapter 2 for a more detailed account of the protest.

5 Other points in the community were under RCMP occupation, as they set up command posts to watch the native protestors.

6 This is a pseudonym.

7 As outlined in chapter 2, the agreement provided boats, licences, money, and training to the Burnt Church First Nation in exchange for their agreement to fish under the regulation of the Canadian government and the DFO.

8 For a detailed discussion of Mi'kmaw engagement with Catholicism in the sixteenth century see Henderson (1997).

9 Here, as in other places, I must point out that there are certainly more responses than these; but among those whom I knew in the community, these are common and significant positions. I look forward to the time when more voices are added from within the community itself about these times and experiences.

10 This work for healing is fundamentally important within Mi'kma'ki, but it also manifests itself as education and activism elsewhere. Among traditionalist activists from Burnt Church, for example, some continue their work within the community, while others displaced in the aftermath of the dispute engage in the same work in other places across North America.

11 This Bible study group was the largest and most stable of a network of home-based Bible studies in the community during my time there. In this particular group, the 6–15 members attended Roman Catholic and Pentecostal congregations and some had begun their own home and Internet-based ministry. The leaders are a retired white couple, Steve and Ann (these are pseudonyms), who attend a Pentecostal Assembly further "up-river" and travel over an hour every week to lead this group. The meeting always begins with chatting and laughter as people arrive, usually includes singing for at least half an hour, and then moves into the teaching time – focusing on Bible passages – led by Steve. Finally, the group moves into prayer, where the needs of people in the group and in the community are raised. Prayer happens with the laying on of hands and sometimes with speaking in tongues. After the formal part of the meeting, a lunch (a Miramichi tradition for all groups) is set out – sandwiches, chips, sweets, and coffee. The lunch is the time for deeper visiting, sometimes for further prayer or theological discussion, and for people to fortify themselves with laughter and calories for the drive home. When I returned to Esgenoôpetitj in 2007 and 2009, this group was still meeting, though leadership of it has been taken on by people from *within* the reserve, and the location of the meetings had shifted to the band council office.

12 Lloyd and Leo both consulted with me on a paper about these issues, and these characterizations arise from our resulting conversations and from Lloyd's response to a draft of this section of the book.

13 Lloyd talks about the ways he and his family are threatened by the evil of the world (violence, drugs, poverty); early in our relationship he spoke to me about the guardian spirits that walk with him and protect him and his family. He has described these spirits to me as traditional Mi'kmaw guardian spirits, and as the angels of God who, through prayer, gather around him, his family, and his home to protect them from evil. Lloyd believes that the Indian Act governments (chiefs, band councils, etc.) were created to teach the people some lesson, which they still have to learn. When they learn this lesson they will be liberated from the corruption of the Indian Act system and its chiefs, since the Creator's purpose will be fulfilled.

14 Author Unknown. I learned this song while participating in the Bible study group.

15 Leo was the leader of the Esgenoôpetitj Rangers.

16 It is interesting to chart these alliances with Christian groups through the dispute, depending on the needs of the community. The CPT was important at one time – but they didn't remain in the community after the dispute. The people who are still around are from the charismatic and Catholic communities; the radical, left-leaning Protestants are mostly absent.

17 This is the usual system in the commercial season. Fishers come in with their lobster catch and buyers or their agents are set up on the wharf to receive and purchase the catch each day. The buyers sell the catch into the processing (canning) or retail food systems.

18 This problem is further complicated by the importance of Conservative and Liberal party politics in the local area and the lack of an effective municipal level of government. While the region has elected MPs and MLAs, there is no local municipal council, like a village or a township. The maintenance and practical needs of the community are the job of the region, but the region does not function as an active elected government.

19 In 1999, out of a total of 307 seats in the House of Commons, New Brunswick held 10.

20 This is a pseudonym.

21 This is a pseudonym.

22 These concerns connect very specifically to concerns about resource conservation, and the ability of residents to make a living in their traditional occupations, fishing and forestry. This will be explored in depth in the next chapter.

23 This critique of Canadian absence was focused on the government, as indicated here, but also included critique of neighbouring communities, as discussed in chapter 5, and of Canadian activists, elaborated in chapter 6.

24 Many Mi'kmaq also have served in the U.S. armed forces, and their names
    are not on this monument to Canadian veterans.

5. Conservation Talk

1   For an example of these parameters, see the terms of appointment of the
    Miramichi Bay Community Relations Panel, as outlined in their final
    report, where "the Panel's mandate specifically excluded any dealings
    with aboriginal rights" (Augustine and Richard 2002b, 7).
2   See Adlam 1999, Paul 2000, and Upton 1979 for discussions of this process
    in Mi'kmaw communities.
3   During the year I was in Burnt Church, the UPM Kymmene mill was on
    strike for six months, a long and bitter dispute. The lack of regional
    support for the millworkers was conspicuous. Mill jobs were the best in
    the region, and locals say it was hard to get in at the mill unless you had a
    family member already working there. During the strike, millworkers left
    their families behind in the Miramichi while they travelled to paying
    work in Alberta's oil sands and in Newfoundland shipyards. The strike
    ended in the fall of 2005, but by 2007 the mill was closed for good. The
    other mill in Miramichi, owned by Weyerhauser, was closed and put up
    for sale in 2006. The exodus of men to the Alberta oil sands for
    employment continues.
4   Both the DFO and many within the First Nation argue that the shifting of
    licences is insignificant, because the overall numbers of boats and traps in
    the fishery remain unchanged.
5   These prices are as found on Tri Nav Marine Brokerage in August 2007.
    TriNav hosts an online listing service for licences and boats across Atlantic
    Canada; see www.trinav.com.
6   Many English residents do not believe that natives were denied access to
    the commercial fishery. Some of the history of the Mi'kmaw fishery at
    Esgenoôpetitj is outlined in chapter 3. Adlam (1999) documents some
    native stories of this denial of access in the Mi'kmaw riverine fishery on
    the Miramichi.
7   While the Passamaquoddy are not officially recognized by the Govern-
    ment of Canada as a First Nation (all Passamaquoddy reserves are in the
    United States), Passamaquoddy people live in Canada as well as in the
    United States. Some of these non-status Passamaquoddy are allies of the
    Mi'kmaq at Esgenoôpetitj, who recognize them as holders of Aboriginal
    and treaty rights. The Abenaki, Mi'kmaq, Maliseet, and Passamaquoddy
    were all signatories to the 1725/26 treaties.

8 The plan cites the collapse of the cod and salmon fisheries, as well as the impending collapse of the snow crab fishery, as evidence of this. (Ward and Augustine 2000a, VII)

9 Within the English community, residents certainly do not think that their concerns about conservation have been manufactured by the DFO and the federal government, as demonstrated by the earlier discussion of conservation and livelihood. Having the government buy the fishing licences in the community in order to retire them, as suggested by the EFN Management Plan, puts dollars into the community but does nothing to solve the ongoing problem of livelihood for the English residents. (This, of course, is the same problem in the First Nation, where government dollars do not necessarily or perhaps even usually translate into sustainable jobs.)

10 Lloyd says: "Our feeling was that [the Canadian government] never read it, never read it, never read it, and we told them, 'We sent it to you.' And they checked and said, 'Well, we never received it' ... I think we faxed it directly to Dhaliwal."

11 These experiences were related to me in interviews with people from Esgenoôpetitj, documented by members of the CPT in their report *Gunboat Diplomacy* (2001), and also related in interviews with ARC-A.

12 It is not clear that such activities are a significant source of overfishing, given the small and local scale on which they happen, in comparison to the overfishing of the corporate participants in the industry (through the regulation of the commons) discussed above.

6. The Canadian Way

1 This provincial policing contract does not preclude New Brunswick municipalities from having their own municipal forces, if they so choose.

2 Eventually, the federal government launched a long-term strategy to deal with the implications of the *Marshall* decision, in which the DFO negotiated fishery access and Indian Affairs renegotiated treaty rights in the Maritimes. This became known as the Molloy process, after its lead negotiator (DFO 2000c). As one reviewer of this book pointed out, New Brunswick First Nations didn't engage with the Molloy process – it "didn't take." Perhaps this is a result of the tension, in Aboriginal New Brunswick politics, between sovereigntists and Indian Act politicians. We see this in Burnt Church, in Tobique, and in Mawiw, for example. The importance of these dynamics is a question worth further exploration and research.

3 Mr Justice Richard, an Acadian New Brunswicker, is the former chief justice of the Court of Queen's Bench of New Brunswick. Chief Augustine

is the former chief of Eel Ground First Nation and also a co-founder of the Atlantic Policy Congress (APC) of First Nation Chiefs.

4 This phrase is taken from Heidegger 1959, 46.

5 John Ralston Saul provides an interesting and controversial recent example of engagement with "peace, order and good government" in *A Fair Country* (2008).

6 Some days after the encounter on the wharf, as he prepared to leave the community of Esgenoôpetitj, Finlay's reflections on his own political position return to this subject: "Have these people (the Mi'kmaq) converted me to their cause? Am I becoming an "Indian lover"? Do I care more about them than the non-native fishers? No, no, and *not really*" (2000, 8; emphasis added).

Postscript

1 A fishery for traditional and ceremonial purposes, as in *Sparrow*.

Bibliography

Adlam, Robert G., Wendy Burnett, and Robert Adlam, eds. 1999. "The Discourse of an Aboriginal Fishery." In *Language and Identity: Papers from the Twenty-third Meeting of the Atlantic Provinces Linguistic Association*, ed. Wendy Burnett and Robert Adlam. Sackville: Mount Allison University.

Alfred, Taiaiake. 2005. *Wasáse: Indigenous Pathways of Action and Freedom.* Peterborough: Broadview Press.

Armitage, Peter, and Daniel Ashini. 1998. "Partners in the Present to Safeguard the Past: Building Cooperative Relations between the Innu and Archaeologists regarding Archaeological Research in Innu Territory." *Études/Inuit/Studies* 22 (2): 31–40.

Augustine, Chief Roger J., and Mr. Justice Guy Richard. 2002a. "Executive Summary." *Miramichi Bay Community Relations Panel. Building Bridges: Miramichi Fishing Communities at a Crossroad.* Department of Fisheries and Oceans, Government of Canada.

Augustine, Chief Roger J., and Mr. Justice Guy Richard. 2002b. *Miramichi Bay Community Relations Panel. Building Bridges: Miramichi Fishing Communities at a Crossroad.* Department of Fisheries and Oceans, Government of Canada.

Basque, Maurice. 1991. *Entre Baie et Péninsule: Histoire de Néguac.* Village of Neguac.

Basso, Keith H. 1996. *Wisdom Sits in Places: Landscape and Language among the Western Apache.* Albuquerque: University of New Mexico Press.

Battiste, Marie. 1997. "Nikanikinútmaqn." In Henderson, *The Míkmaw Concordat*, 13–20.

Bélanger, Claude. 2001. *Peace, Order and Good Government.* Montreal: Marianopolis College. http://faculty.marianopolis.edu/c.belanger/ quebechistory/federal/pogg.htm.

Bellah, Robert N. 1967. "Civil Religion in America." *Daedalus, Journal of the American Academy of Arts and Sciences* 96 (1): 1–21.

Bellah, Robert N. 1991. "Introduction: Civil Religion in America." In *Beyond Belief: Essays on Religion in a Post-Traditionalist World*. Berkeley: University of California Press.

Berger, Thomas. 1977. *Northern Frontier, Northern Homeland: The Report of the Mackenzie Valley Pipeline Inquiry*. Ottawa: J. Lorimer in association with Publishing Centre, Supply and Services Canada.

Berglund, Eeva, and David G. Anderson. 2003. "Introduction: Towards an Ethnography of Ecological Underprivilege." In *Ethnographies of Conservation*, ed. Anderson and Berglund, 1–18. New York: Berghahn Books.

Borrows, John. 2009. *Canada's Indigenous Constitution*. Toronto: University of Toronto Press.

Brown, Charles. 2003. "The Real and the Good: Phenomenology and the Possibility of an Axiological Rationality." In *Eco-Phenomenology: Back to the Earth Itself*, ed. Charles Brown and Ted Toadvine, 3–18. Albany: State University of New York Press.

Bruner, Edward M. 1997. "Ethnography as Narrative." In *Memory, Identity, Community: The Idea of Narrative in the Human Sciences*, ed. Lewis P. and Sandra K. Hinchman, 264–80. Albany: State University of New York.

Caddy, James. 2001. "The Caddy Report: The Lobster Resource in the Miramichi Bay." Department of Fisheries and Oceans, Government of Canada.

Cameron, Alex M. 2009. *Power without Law: The Supreme Court of Canada, the Marshall Decisions, and the Failure of Judicial Activism*. Montreal: McGill-Queen's University Press.

Camp, Dalton. 1999. "Silence Is Not Good Enough." *New Brunswick Telegraph Journal*, 5 October 1999: A7.

Canadian Press. 1999. "Natives to Observe Commercial Season." 22 October 1999.

Canadian Press. 2000. "Tensions Heating Up in Native Fishery." 6 May 2000.

Capps, Walter. 1995. *Religious Studies: The Making of a Discipline*. Minneapolis: Augsburg Fortress.

Casey, Edward S. 1993. *Getting Back into Place: Toward a Renewed Understanding of the Place-World*. Bloomington: Indiana University Press.

CBC. 1999a. "Court Limits Marshall Decision." *CBC News Indepth: Fishing Fury*, 17 November 1999.

CBC. 1999b. "First Nations React to Marshall Clarification." *CBC News Indepth: Fishing Fury*, 18 November 1999.

CBC. 2000. "N.B. Natives and DFO Face Off over Lobster." *CBC News Indepth: Fishing Fury*, 12 August 2000.

CBC. 2002. "Panel Aims for Peace in Burnt Church Dispute." *CBC News Indepth: Fishing Fury*, 9 April 2002.

CBC. 2010. "7 N.B. Communities among Canada's Poorest." *CBC News – New Brunswick*, 23 February 2010.

Chawla, Louise. 1994. *In the First Country of Places: Nature, Poetry and Childhood Memory*. New York: State University of New York Press.

Cheney, Jim. 1997. "Postmodern Environmental Ethics: Ethics as Bioregional Narrative." In *Memory, Identity, Community: The Idea of Narrative in the Human Sciences*, ed. Lewis P. and Sandra K. Hinchman, 328–59. Albany: State University of New York Press.

Christian Peacemaker Teams. 2001. *Gunboat Diplomacy: Canada's Abuse of Human Rights at Esgenoôpetitj (Burnt Church, New Brunswick)*. Toronto: Christian Peacemaker Teams.

Coates, Ken. 2000. *The Marshall Decision and Native Rights*. Montreal: McGill-Queen's University Press.

Conklin, Beth A. 1997. "Body Paint, Feathers and VCRs: Aesthetics and Authenticity in Amazonian Activism." *American Ethnologist* 24 (4): 711–37.

Conklin, Beth A. 2002. "Shamans versus Pirates in the Amazonian Treasure Chest." *American Anthropologist* 104 (4): 1050–61.

Conklin, Beth A., and Laura R. Graham. 1995. "The Shifting Middle Ground: Amazonian Indians and Eco-Politics." *American Anthropologist* 97 (4): 695–710.

Creswell, John W. 1998. *Qualitative Inquiry and Research Design: Choosing among Five Traditions*. Thousand Oaks: Sage.

Crosby, Alfred W. 1972. *The Columbian Exchange: Biological and Cultural Consequences of 1492*. Westport: Greenwood Press.

Crosby, Alfred W. 1986. *Ecological Imperialism: The Biological Expansion of Europe, 900–1900*. New York: Cambridge University Press.

Deloria Jr, Vine. 2003. *God Is Red: A Native View of Religion*. 3rd ed. Golden: Fulcrum Publishing.

Department of Fisheries and Oceans. 1999. "Statement by Herb Dhaliwal, Minister of Fisheries and Oceans. Update on Marshall Case Ruling." 1 October 1999. Government of Canada.

Department of Fisheries and Oceans. 2000a. *Infocéan, Juin 2000*. Government of Canada.

Department of Fisheries and Oceans. 2000b. "Letter to the Editor." (Responding to John Cummins MP, signed by Fisheries Minister Herb Dhaliwal.) Government of Canada.

Department of Fisheries and Oceans. 2000c. "Statement by Herb Dhaliwal, Minister of Fisheries and Oceans. Update on Fisheries Affected by the Supreme Court's Marshall Decision." 21 September 2000. Government of Canada.

Department of Fisheries and Oceans. 2001. "Ministers Announce Negotiators, Process for Long-term Response to Marshall." Government of Canada.

Department of Fisheries and Oceans. 2002. "Key Elements of an Agreement-in-Principle for a Comprehensive Fisheries Agreement with Burnt Church." Government of Canada.

Department of Justice, Canada. 1982. *Canadian Charter of Rights and Freedoms.* Schedule B: Constitution Act 1982. Government of Canada.

Devall, Bill, and George Sessions. 1985. *Deep Ecology: Living as if Nature Mattered.* Salt Lake City: Gibbs Smith.

Dharamsi, Titch. 2000. "No Retreat from Burnt Church." *National Post,* 20 September 2000.

Dombrowski, Kirk. 2002. "The Praxis of Indigenism and Alaska Native Timber Politics." *American Anthropologist* 104 (4): 1062–73.

Doyle-Bedwell, Patricia, and Fay G. Cohen. 2001. "Aboriginal Peoples in Canada: Their Role in Shaping Environmental Trends in the Twenty-first Century." In *Governing the Environment: Persistent Challenges, Uncertain Innovations,* ed. Edward A. Parson, 169–206. Toronto: University of Toronto Press.

Dyck, Noel. 1991. *What Is the Indian "Problem"? Tutelage and Resistance in Canadian Indian Administration.* St John's: Institute of Social and Economic Research, Memorial University of Newfoundland.

Finlay, John. 2000. "Burnt Church Diaries." Canadian Broadcasting Corporation.

Foucault, Michel. 2007. *Security, Territory, Population: Lectures at the Collège De France 1977–78.* New York: Palgrave MacMillan.

Francis, Daniel. 1992. *The Imaginary Indian: The Image of the Indian in Canadian Culture.* Vancouver: Arsenal Pulp Press.

Freeman, Victoria. 2002. *Distant Relations: How My Ancestors Colonized North America.* Toronto: Random House.

Frodeman, Robert. 2005. "Place." In *Encyclopedia of Science, Technology and Ethics,* ed. Carl Mitcham, 1409–12. Detroit: Thomson Gale.

Gandhi, Leela. 1998. *Postcolonial Theory: A Critical Introduction.* New York: Columbia University Press.

Geertz, Clifford. 1966. "Religion as a Cultural System." In *Anthropological Approaches to the Study of Religion,* ed. M. Banton, 1–46. New York: Praeger.

gkisedtanamoogk. 2007. Personal communication (email), 9 August 2007.

Greer, Allan. 2000. *The Jesuit Relations: Natives and Missionaries in Seventeenth-Century North America.* Boston: Bedford/St Martin's.

Greer, Allan. 2005. *Mohawk Saint: Catherine Tekakwitha and the Jesuits.* New York: Oxford University Press.

Guha, Ramachandra. 1989a. "Radical Environmentalism: A Third World Critique." *Environmental Ethics* 11 (1): 71–83.

Guha, Ramachandra. 1989b. *The Unquiet Woods: Ecological Change and Peasant Resistance in the Himalaya*. Delhi: Oxford University Press.

Gupta, Akhil, and James Ferguson. 1997. *Culture, Power, Place: Explorations in Critical Anthropology*. Durham: Duke University Press.

Hebert, Chantal. 1999. "Trudeau Started This Lobster War." *New Brunswick Telegraph Journal*, 7 October 1999: A7.

Heidegger, Martin. 1959. *Discourse on Thinking*, trans. John M. Anderson and E. Hans Freund. New York: Harper and Row Publishers.

Heidegger, Martin. 1962. *Being and Time*, trans. John Macquarrie and Edward Robinson. London: SCM Press.

Henderson, James (Sakej) Youngblood. 1997. *The Mikmaw Concordat*. Halifax: Fernwood Publishing.

Henderson, James (Sakej)Youngblood. 2006. *First Nations Jurisprudence and Aboriginal Rights: Defining the Just Society*. Saskatoon: Native Law Centre, University of Saskatchewan.

Henderson, James (Sakej) Youngblood. 2007. *Treaty Rights in the Constitution of Canada*. Scarborough: Carswell.

Hornborg, A. 1998. "Mi'kmaq Environmentalism: Local Incentives and Global Projections." In *Sustainability – The Challenge: People, Power and the Environment*, ed. A. Sandberg and S. Sörlin, 202–11. Montreal: Black Rose Books.

Ingold, Tim. 2000. *The Perception of the Environment: Essays in Livelihood, Dwelling and Skill*. New York: Routledge.

Isaac, Thomas. 2001. *Aboriginal and Treaty Rights in the Maritimes: The Marshall Decision and Beyond*. Saskatoon: Purich Publishing Limited.

King, Roger J.H. 1999. "Narrative, Imagination and the Search for Intelligibility in Environmental Ethics." *Ethics and the Environment* 4 (1): 23–38.

King, Sarah. 2011. "Conservation Controversy: Sparrow, Marshall, and the Mi'kmaq of Esgenoôpetitj." *International Indigenous Policy Journal* 2 (4). http://ir.lib.uwo.ca/iipj/vol2/iss4/5.

King, Thomas. 2003. *The Truth about Stories: A Native Narrative*. Toronto: House of Anansi Press.

Klager, Bob. 2002. "A Nation Divided: Despite Outward Signs of Peace, Burnt Church Still Facing Internal Turmoil." *Saint John Telegraph Journal*, 24 August 2002.

Kovach, Margaret. 2009. *Indigenous Methodologies: Characteristics, Conversations and Contexts*. Toronto: University of Toronto Press.

Kretch, Shepard, III. 1999. *The Ecological Indian: Myth and History* New York: W.W. Norton & Company.

Kunstler, James Howard. 1993. *The Geography of Nowhere: The Rise and Decline of America's Man-Made Landscape*. New York: Simon and Schuster.

Linden, Sidney B. 2007. *The Report of the Ipperwash Inquiry*. Ministry of the Attorney General, Government of Ontario.

Low, Setha M., and Denise Lawrence-Zuniga. 2003. *The Anthropology of Space and Place: Locating Culture*. Malden: Blackwell.

MacIntyre, Alisdair. 1997. "The Virtues, the Unity of a Human Life, and the Concept of a Tradition." In *Memory, Identity, Community: The Idea of Narrative in the Human Sciences*, ed. Lewis P. and Sandra K. Hinchman, 214–63. Albany: State University of New York Press.

Macy, Joanna. 1995. "The Ecological Self: Postmodern Ground for Right Action." In *Readings in Ecology and Feminist Theology*, ed. Mary Heather MacKinnon and Moni MacIntyre, 259–69. Franklin, WI: Sheed and Ward.

Malpas, J.E. 1999. *Place and Experience: A Philosophical Topography*. New York: Cambridge University Press.

Malpas, J.E. 2006. *Heidegger's Topology: Being, Place, World*. Cambridge: MIT Press.

Maracle, Lee, Jeannette C. Armstrong, Delphine Derickson, and Greg Young-Ing. 1993. *We Get Our Living like Milk from the Land*. Penticton: Theytus Books.

Marietta, Don. 2003. "Back to Earth with Reflection and Ecology." In *Eco-Phenomenology: Back to the Earth Itself*, ed. Charles Brown and Ted Toadvine. Albany: State University of New York Press.

Martin, Barb. N.d. "History of the Mi'kmaq and the Burnt Church First Nation." Unpublished paper.

McMillan, Leslie Jane. 2002. "Koqqwaja'ltimk: Mi'kmaq Legal Conscious-ness." PhD dissertation, University of British Columbia. National Library of Canada, http://www.collectionscanada.gc.ca/obj/s4/f2/dsk4/etd/NQ79241.PDF.

McMullan, John L., and David C. Perrier. 1997. "Poaching vs. the Law: The Social Organization of Illegal Fishing." In *Crimes, Laws and Communities*, ed. John L. McMullan, David C. Perrier, Stephen Smith, and Peter D. Swan, 29–60. Halifax: Fernwood Publishing.

Memmi, Albert. 1965. *The Colonizer and the Colonized*. New York: Orion Press.

Menzies, Charles. 1994. "Stories from Home: First Nations, Land Claims, and Euro-Canadians." *American Ethnologist* 21 (4): 776–91.

Miller, Virgina P. 1976. "Aboriginal Micmac Population: A Review of the Evidence." *Ethnohistory* (Columbus, OH) 23 (2): 117–28.

Mitchell, Timothy. 1991. *Colonising Egypt*. Berkeley: University of California Press.

Mofina, Rick. 1999. "Aboriginal Fishermen Reject Chiefs' Plea: The Call for a Self-Imposed Moratorium on Fishing Has Been Rejected because 'to Do Anything Else Would Be Criminal,' Burnt Church First Nation Controller Says." *Montreal Gazette*, 8 October 1999.

Morris, Chris. 1999. "Chiefs Tell Natives to Keep On Fishing." *Moncton Times & Transcript*, 30 September 1999: A4.

Mugerauer, Robert. 1994. *Interpretations on Behalf of Place: Environmental Displacements and Alternative Responses*. Albany: State University of New York Press.

Mugerauer, Robert. 2008. *Heidegger and Homecoming: The Leitmotif in the Later Writings*. Toronto: University of Toronto Press.

Mulhall, Stephen. 2005. *Heidegger and Being and Time*. 2nd ed. New York: Routledge.

Nash, Ronald J., and Virginia P. Miller. 1987. "Model Building and the Case of the Micmac Economy." *Man in the Northeast* 34: 41–56.

Notzke, Claudia. 1994. *Aboriginal Peoples and Natural Resources in Canada*. North York: Captus University Press.

Nye, Malory. 2008. *Religion: The Basics*. 2nd ed. New York: Routledge.

Obomsawin, Alanis. 2002. *Is the Crown at War with Us?* Documentary film. National Film Board of Canada.

Orsi, Robert. 2002. *The Madonna of 115th Street: Faith and Community in Italian Harlem, 1880–1950*. New York: Yale University Press.

Paul, Daniel N. 2000. *We Were Not the Savages: A Mi'kmaq Perspective on the Collision between European and Native American Civilizations*. Halifax: Fernwood.

Peña, Devan C. 2002. "Endangered Landscapes and Disappearing Peoples: Identity, Place and Community in Ecological Politics." In *The Environmental Justice Reader: Politics, Poetics and Pedagogy*, ed. Joni Adamson, MeiMei Evans, and Rachel Stein, 58–81. Tucson: University of Arizona Press.

Perley, M.H. 1841. "Extract from a Report to His Excellency the Lieutenant-Governor of New-Brunswick, by M.H. Perley, Commissioner for Indian Affairs, dated 11th December 1841." In *The Native Peoples of Atlantic Canada: A History of Indian-European Relations*, ed. H.F. McGee, 81–9. Don Mills: Oxford University Press, 1983.

Poitras, Jacques. 1999a. "Dhaliwal Hints He May Impose Rules: Fisheries Minister Promises 'Final Decisions' within Days." *New Brunswick Telegraph Journal*, 6 October 1999: A1–2.

Poitras, Jacques. 1999b. "Natives Won't Turn the Other Cheek: Mi'kmaq Fishermen Say They Can't Wait until Spring when Non-natives Go Lobster Fishing." *New Brunswick Telegraph Journal*, 5 October 1999: A3.

Porteous, J. Douglas, and Sandra E. Smith. 2001. *Domicide: The Global Destruction of Home*. Montreal: McGill-Queen's University Press.

Porter, Tim. 1999. "Minister Allows Native Fishery." *Moncton Times & Transcript*, 2 October 1999: C1–2.

Prins, Harald. 1996. *The Mi'kmaq: Resistance, Accommodation, and Cultural Survival*. Toronto: Harcourt Brace College Publishers.

R. v. Marshall. 1999a. Supreme Court of Canada (Docket 26014), 17 September 1999.

R. v. Marshall. 1999b. Supreme Court of Canada (Docket 26014), 17 November 1999.

R. v. Sparrow. 1999. Supreme Court of Canada (Docket 20311), 31 May 1999.

Relph, Edward. 2008. "Disclosing the Ontological Depth of Place: *Heidegger's Topology* by Jeff Malpas." *Environmental and Architectural Phenomenology Newsletter* 19 (1): 5–8. www.arch.ksu.edu/seamon/Relph_Malpasreview08. htm.

Richer, Stephen. 1988. "Fieldwork and the Commodification of Culture: Why the Natives Are Restless." *Canadian Review of Sociology and Anthropology / La Revue Canadienne de Sociologie et d'Anthropologie* 25 (3): 406–20.

Robbins, Joel. 2004a. *Becoming Sinners: Christianity and Moral Torment in Papua New Guinea Society*. Berkeley: University of California Press.

Roepstorff, Andreas, and Nils Bubandt. 2003. "General Introduction: The Critique of Culture and the Plurality of Nature." In Roepstorff et al., *Imagining Nature*, 9–28.

Roepstorff, Andreas, Nils Bubandt, and Kalevi Kull, eds. 2003. *Imagining Nature: Practices of Cosmology and Identity*. Denmark: Aarhus University Press.

Royal Commission on Aboriginal Peoples. 1996. *Royal Commission Report on Aboriginal Peoples*. Indian and Northern Affairs Canada.

Saul, John Ralston. 2008. *A Fair Country: Telling Truths about Canada*. Toronto: Penguin.

Smith, David M. 2000. *Moral Geographies: Ethics in a World of Difference*. Edinburgh: Edinburgh University Press.

Smith, Mick. 2001. *An Ethics of Place: Radical Ecology, Postmodernity and Social Theory*. Albany: State University of New York Press.

Somerville, Fenton. 2000. "Burnt Church Diaries." *CBC News Indepth: Fishing Fury*, 14 September 2000.

Srivastava, Sarita. 2005. "'You're Calling Me a Racist?' The Moral and Emotional Regulation of Antiracism and Feminism." *Signs: Journal of Women in Culture and Society* 31 (1): 29–62.

Stefanovic, Ingrid Leman. 1994. "What Is Phenomenology?" *Brock Review* 3 (1): 58–77.

Stefanovic, Ingrid Leman. 2000. *Safeguarding Our Common Future: Rethinking Sustainable Development*. Albany: State University of New York Press.

Trigger, Bruce. 1985. *Natives and Newcomers: Canada's "Heroic Age" Revisited*. Montreal: McGill-Queen's University Press.

Trigger, Bruce. 1996. *The Cambridge History of the Native Peoples of the Americas*, vol. 1, *North America*. New York: Cambridge University Press.

Trouillot, Michel-Rolph. 1995. *Silencing the Past: Power and the Production of History*. Boston: Beacon Press.

Tuan, Yi-Fu. 1974. *Topophilia: A Study of Environmental Perceptions, Attitudes and Values*. Englewood Cliffs: Prentice-Hall.

United Nations. 1948. *Universal Declaration of Human Rights*. General Assembly Resolution 217 A (III) of 10 December 1948.

Upton, L.F.S. 1979. *Micmacs and Colonists: Indian–White Relations in the Maritimes 1713–1867*. Vancouver: UBC Press.

Vandergeest, Peter, and E. Melanie DuPuis. 1996. "Introduction." In *Creating the Countryside: The Politics of Rural and Environmental Discourse*, ed. Melanie E. DuPuis and Peter Vandergeest, 1–25. Philadelphia: Temple University Press.

Vickers, Kevin. 2004. "The RCMP and the Canadian Way." *Gazette* 66 (3) Ottawa: Royal Canadian Mounted Police.

Ward, James, and Lloyd Augustine. 2000a. *Draft for EFN [Esgenoôpetitj First Nation] Fishery Act*. Esgenoôpetitj: Esgenoôpetitj First Nation.

Ward, James, and Lloyd Augustine. 2000b. *Draft for EFN Management Plan*. Esgenoôpetitj: Esgenoôpetitj First Nation.

Warren, Kay. 1998. *Indigenous Movements and Their Critics: Pan-Maya Activism in Guatemala*. Princeton: Princeton University Press.

Wasson, Manford, and Lottie Wasson Murdoch. 1999. *Church History, the Last 100 Years, 1899–1999. 100th Anniversary, St. David's United Church, Burnt Church N.B.* St David's United Church.

Wicken, William C. 2002. *Mi'kmaq Treaties on Trial: History, Land, and Donald Marshall Junior*. Toronto: University of Toronto Press.

Widdowson, Frances, and Albert Howard. 2008. *Disrobing the Aboriginal Industry: The Deception behind Indigenous Cultural Preservation*. Montreal: McGill-Queen's University Press.

Wien, Fred. 1986. *Rebuilding the Economic Base of Indian Communities: The Micmac in Nova Scotia*. Montreal: The Institute for Research on Public Policy.

Wildsmith, Bruce H. 2001. "Vindicating Mi'Kmaq Rights: The Struggle before, during, and after Marshall." *Windsor Yearbook of Access to Justice* 19 (203).

Williams, Daniel R., and Susan I. Stewart. 1998. "Sense of Place: An Elusive Concept That Is Finding a Home in Ecosystem Management." *Journal of Forestry* 96 (5): 18–23.

Index

11 September 2001, 50; impact on dispute, 50; post-9/11 political context, 51; War on Terror, 50

Aboriginal, 9–10, 15, 17 20, 23, 29, 38, 46–7, 62, 66, 69, 79, 82–3, 90, 93, 97, 108, 117–18, 120, 122, 135, 143–4, 146–8, 153–6, 172n–3n; Aboriginal rights, 10, 20, 38, 46, 82, 120, 135, 143, 154, 172n; between Aboriginal and treaty rights, 82; activists and academics, 18; religious and cultural meaning and identity, 82. *See also* Aboriginals; indigenous; native

Aboriginals, 18, 120–1; constitutional rights of, 148; Dineh (Navajo), 117; engaging in research with, 17; inclusion in resource management decisions, 93; regeneration and resurgence of, 122

Aboriginal Affairs Secretariat, 47

Aboriginal communities, 17, 19, 23, 79, 90, 120; discourse about ethics of research within, 23; revitalized, 79

Aboriginal people, 9, 17, 29, 38, 69, 93, 97, 108, 118, 120, 122, 148, 153–5. *See also* Aboriginals

Aboriginal Rights Coalition–Atlantic's Observer Project (ARC–A), 20; accounts of dispute, 154

Acadians, 63–5; alliance with Mi'kmaw people, 64; Equal Opportunity Program, 69; Le Grand Dérangement, 64; Néguac, 62, 65–7, 69, 144, 163, 169n

Acadian Peninsula, 5, 56

Alfred, Taiaiake, 119, 156

Alliance Party, 148; criticism of DFO, 147

Al-Qaeda, 50. *See also* 11 September 2001

American Academy of Religion, 70

anglophone, 6, 138

Annapolis Royal, 83

anthropology, 11, 16, 69, 99, 113; anthropologists of place, dwelling, and livelihood, 69; perception of adoption of Christianity, 101; qualitative research, 14, 16–17, 99, 165n; sharing of stories as "thick